Time's Up!
Mandatory Retirement in Canada

CAUT SERIES TITLES

Time's Up! Mandatory Retirement in Canada, eds. C.T. (Terry) Gillin, David MacGregor, and Thomas R. Klassen (2005)

Disciplining Dissent, eds. William Bruneau and James L. Turk (2004)

Let Them Eat Prozac by David Healy (2003)

Counting Out the Scholars: How Performance Indicators Undermine Colleges and Universities by William Bruneau and Donald C. Savage (2002)

The Olivieri Report: The Complete Text of the Report of the Independent Inquiry Commissioned by the Canadian Association of University Teachers by Jon Thompson, Patricia Baird, and Jocelyn Downie (2001)

The Corporate Campus: Commercialization and the Dangers to Canada's Colleges and Universities, ed. James L. Turk (2000)

Universities for Sale: Resisting Corporate Control over Canadian Higher Education by Neil Tudiver (1999)

Time's Up! Mandatory Retirement in Canada

C.T. (Terry) Gillin
David MacGregor and
Thomas R. Klassen, editors

A CAUT Series Title

James Lorimer & Company Ltd., Publishers

Toronto

James Lorimer & Company Ltd. acknowledges the support of the Ontario Arts Council. We acknowledge the support of the Government of Canada through the Book Publishing Industry Development Program (BPIDP) for our publishing activities. We acknowledge the support of the Canada Council for the Arts for our publishing program. We acknowledge the support of the Government of Ontario through the Ontario Media Development Corporation's Ontario Book Initiative.

Cover design: Meghan Collins

The Canada Council for the Arts | Le Conseil des Arts du Canada

ONTARIO ARTS COUNCIL
CONSEIL DES ARTS DE L'ONTARIO

Library and Archives Canada Cataloguing in Publication

Time's up! : mandatory retirement in Canada / edited by C. T. (Terry) Gillan, David MacGregor and Thomas R. Klassen.

Includes bibliographical references and index.
ISBN 1-55028-883-0

1. Retirement, Mandatory—Canada. 2. Age discrimination in employment—Canada. I. Gillan, C. T. II. MacGregor, David III. Klassen, Thomas

HD7108.4.C3T54 2005 331.3'98 C2005-902786-X

A CAUT Series Title
James Lorimer & Company Ltd,
Publishers
35 Britain Street
Toronto, Ontario
M5A 1R7

www.lorimer.ca

Printed and bound in Canada.

Contents

To all involuntarily retired workers in Canada

Acknowledgements

The editors express special thanks to Robin Haley-Gillin for her editorial assistance, to Rachelle Anne Mendoza for her extensive research and editing support, and to Katherine Bryan for her editing.

We acknowledge the financial assistance from the Restructuring Work and Labour in the New Economy project, funded by the Social Sciences and Humanities Research Council, and directed by the Centre for Research on Work and Society at York University.

We are grateful to William Bruneau and James Turk of the Canadian Association of University Teachers for their encouragement, guidance and assistance. Bill's contributions in particular have helped to shape the book and have significantly improved the original manuscript. We also wish to thank Jean Lawrence for her excellent style editing, Stacey Curtis for her hard work as production editor, and Russell Wodell for his first-rate effort as proofreader, and copy editor. We thank the publisher, James Lorimer, who has chosen to work with the Canadian Association of University Teachers on a series of publications and has included our book in the series.

We appreciate the support of our three universities, Ryerson, King's University College, and York. Our institutions provide us with the opportunity and resources to learn, teach, research, and write. Our colleagues and students provide ongoing challenge and encouragement to articulate our arguments as clearly as we are able.

Thomas Klassen is particularly indebted to colleagues in the Labour Studies Program at York University for stimulating his interest in the Canadian labour movement. He is also appreciative of the support for this project from members of the Department of Political Science at York University.

Most importantly, we are indebted to the authors who contributed chapters to this volume. Their analyses, interpretations, insights, and passion helped us—and we trust will 'also substantially assist readers—to better understand our social reality.

On a more personal note, the editors thank those who are closest to us and who gave us personal support and encouragement throughout the writing and editing processes—Robin and Luke Gillin; Patricia, Ethan, and Rachael Bishop; and Sue Han. From them we learn what is most important in life. To them we offer what is best in ourselves.

The research conducted for Chapter 4 was funded by a grant from the York University Centre for Research on Work and Society as part of its SSHRC Initiative on the New Economy Research Alliance project. Comments received from Bob Baldwin on an earlier draft of this chapter are gratefully acknowledged

The authors of Chapter 7 shared equally in its research and writing. They wish to thank Brenda Morrison for her research assistance.

Parts of Chapter 9 draw on the author's "Mandatory Retirement and Older Workers: Encouraging Longer Working Lives," *Commentary 200*, Toronto: C.D. Howe Institute, June 2004; the author thanks the C.D. Howe Institute for permission to use this material.

1

Introduction: The Context of Mandatory Retirement

C.T. (Terry) Gillin

In 1979, a special committee of the Senate under the chairmanship of David Croll published *Retirement Without Tears,* calling for the systematic but gradual elimination of mandatory retirement.[1] In the same year, James Pesando published *The Elimination of Mandatory Retirement* for the Ontario Economic Council.[2] Pesando's view was that although human rights were the prime mover of the concerns over mandatory retirement, economic analysis emphasized the efficiency of compulsory retirement.[3]

During the following decade, a number of legislative actions altered retirement rules for many Canadians. For example, in 1982, Quebec legislated a ban on mandatory retirement, followed by Manitoba in 1983.[4] New Brunswick also made legal changes in 1982 that permitted the filing of a human rights complaint by those forced into retirement, except in cases where pension plan requirements are in place. The federal government eliminated mandatory retirement in the federal public service in 1986. The 1987 report of the Ontario Task Force on Mandatory Retirement concluded that arbitrary retirement practices were unacceptable, objectionable, and "sadly out of step with the growing concern and recognition of individual rights."[5]

By 1990, legal challenges to mandatory retirement based on the Canadian Charter of Rights and Freedoms had reached the Supreme Court of Canada.[6] The Court was equivocal and cautious in its analysis. In its *McKinney* decision, the Court acknowledged mandatory retirement as discriminatory but found it a reasonable limit on rights, mainly for socio-economic reasons.[7] The impact of this decision was to curtail further public initiatives for almost a decade.

Overall, mandatory retirement has attracted moderate attention from researchers in Canada since the landmark Croll Senate report of 1979. A review of scholarly research shows a fairly even pattern of work during the two decades beginning with 1979 and an increased rate of publication since 2000. Following the Croll and Pesando reports of 1979, Morley Gunderson and Pesando published together on the topic in 1980.[8] Subsequently, their articles and public policy institute papers emerged with regularity and are central to the discussion and debate. Other researchers and analysts have followed, sometimes supporting, sometimes challenging their work.[9] The output of scholarship since 1979, however, remains modest and dominated by a few authors.[10] As such, this book represents the first systematic, albeit incomplete, examination of mandatory retirement in Canada. The effort is particularly relevant at a time when public debate on the issue has arisen, to which this book seeks to contribute.

More recently, the public voices of government departments and commissions have again questioned the fairness of forced retirement. In 1999, the Department of Justice commissioned a panel to review federal human rights legislation. The Canadian Human Rights Commission welcomed the panel's report, *Promoting Equality*,[11] which included the recommendation to study the issue of mandatory retirement and to eliminate "blanket defences" of it.[12] In 2001, the Ontario Human Rights Commission issued a report, *Time for Action*, which recommended changes to Ontario's human rights code that would essentially eliminate compulsory retirement at age sixty-five and would extend protection to workers who were sixty-five and older.[13] Although no legislative changes have yet resulted from the two human rights reports, a public political debate on mandatory retirement has become more evident. Also in 2001, the British Columbia Court of

Appeal called on the Supreme Court of Canada to reconsider its *McKinney* decision.[14]

Political and other public figures have recently questioned the utility of mandatory retirement. In early 2003, Ontario's Progressive Conservative government introduced legislation designed to eliminate mandatory retirement.[15] The bill "died on the order paper" when the Progressive Conservatives called an election and subsequently lost. The Liberal government of Premier Dalton McGuinty, which won the 2004 Ontario election, has stated its intentions to eliminate mandatory retirement.[16] Upon attaining office in late 2003, Prime Minister Paul Martin asserted, "I don't believe in mandatory retirement," and called for public debate.[17] One quick response came from the president of the Canadian Auto Workers, who defended mandatory retirement as providing "orderly employee turnover," job opportunities for younger workers, and "a certain degree of job security for younger workers, especially during periods of economic downturn."[18] In April 2004, Bank of Canada Governor David Dodge referred to mandatory retirement as a "silly policy; I think any organization is not terribly well served by it."[19] In August 2004, the Ontario Ministry of Labour announced that the McGuinty government is committed to ending mandatory retirement and initiated a consultation process.[20] Precisely because energetic public discussion has begun again, it is well to draw together the latest analyses. It is time that scholars catch up with the public rhetoric in critical studies of the matter.

This review of public discussion identifies two key themes in the debate: human rights and economic utility. However distinctive individual cases may be, the key controversies include the nature of human rights and age discrimination, the role of the state in regulating employment, and the contested arena of the workplace, in part focussing on retirement policies and pension plans. Early in one's career, the issue of retirement seems to be a low priority. For those who enter the workforce later, or whose careers are interrupted by family or other responsibilities, the question of a reasonable pension is likely to present itself sooner rather than later. Moreover, everyone's life circumstances, attitudes, and concerns evolve over time.

Other latent themes are present. The indifference toward ageist attitudes and structures is itself a source of profound unease. Ageism[21] is not essentially different from racism, sexism, homophobia, and other forms of discrimination. Those who experience arbitrary discrimination are denied fair opportunities and are demeaned. Although people generally express many positive attitudes toward aging, all adult age groups, including older adults themselves, share a number of negative stereotypes of the old and are likely to discriminate against them.[22] Stereotypes of older people in employment include assumptions that they are less educated and less motivated than younger people, and that aging necessarily results in a decline in intelligence and ability, lower work performance, poor adaptability to change, and less capacity for retraining.[23] The most pernicious aspect of ageism is that it becomes *systemic*, built into the organization of our lives and taken for granted, *as if* these social arrangements were natural. To deny employment, promotion, income, benefits, or other opportunities on the basis of race, gender, or religion is institutionally unacceptable, though once—not so long ago—these were commonplace. Systemic ageism, however, continues to affect older workers' life chances.

Another theme is the relationship between personal experience and the institutional organization of our lives.[24] All the authors in this volume—though some more explicitly than others—write with passion combined with insight. They write out of their life experiences, educated by their accomplishments, frustrations, and ambivalences. Demographic, legal, and economic debates are the forms into which are poured reflections on life experiences.

The demographic structure of Canada means that an unprecedented number of workers will reach retirement age relatively soon, and under current conditions more will *experience* the policies of forced retirement. Under the Charter and the evolution of human rights legislation, Canada has an expanding consciousness of rights and freedom. Through these political and structural changes, we can extend the personal productivity of many who would otherwise be held back by prejudice, including older people. Equality protection needs to be equally available to all. Immigrants, women, and unskilled labourers are the most vulnerable to age discrimination, and age overlaps in significant ways with other personal characteristics, such as gender and (dis)ability. The terms and conditions of employment, the organization of professional progress, and any available benefits affect our lives in profoundly personal ways. Some find motivation in the desire simply to continue being employed at something they love.

The debate will continue with policy and structural changes emerging over time. Even if the current legal policies supporting mandatory retirement at age sixty-five were to be completely eliminated in the near future, individual and family needs and resources (primarily health and income), as well as pension provisions and bona fide occupational requirements, would continue to be the major influences on individual retirement dates. The outcomes in large measure depend on the intensity and quality of public discussion. This volume aims to inspire and contribute to the debate.

In this volume, a variety of broadly conceived, interrelated arguments frame the analysis of mandatory retirement: theoretical (post-modernist), demographic, social structural, political, economic, comparative, and legal. Within these overlapping perspectives, researchers and analysts develop, elaborate, and test their ideas. Virtually all the authors draw from and rely upon more than one approach.

David MacGregor (Chapter 2) reminds us that the current debate over mandatory retirement arises predominantly out of the experience and reflection of the academic community. Academics have been the most active in questioning the rationale for retirement at a specified age, as illustrated by some of the key cases that have been heard by the Supreme Court of Canada. He notes that university policies and practices regarding forced retirement reflect societal and institutional power structures, and he argues that in the Supreme Court's decisions, universities and their faculty members were stand-ins for employers and employees generally. Mandatory retirement, as a seemingly age-neutral employment strategy, is analyzed as structural inequality. Based upon the available evidence, fears that eliminating arbitrary retirement will lead to a surfeit of older faculty are unfounded; in any case, older faculty who stay past age sixty-five (where this is possible) are some of the most committed teachers and most productive researchers.

C.T. (Terry) Gillin and Thomas R. Klassen (Chapter 3) analyze the

1990 *McKinney* and other pertinent judicial decisions in order to examine the changing legal context of analysis of forced retirement. Virtually every analysis of this issue has been informed by and, at least implicitly, written in response to the *McKinney* decision. Gillin and Klassen argue that the shifting judicial foundation reveals at least two major changes in direction in the past twenty-five years. In the 1980s, the Supreme Court of Canada seemed to stake out a critical position on ageism and mandatory retirement. Subsequently, throughout most of the 1990s, the Supreme Court acknowledged mandatory retirement as a form of discrimination but rationalized it as constitutionally justified, relying predominantly on socio-economic considerations. More recently, since 1999, the Court seems to emphasize the need to assure human dignity, echoing human rights principles, and thereby potentially providing a judicial basis for a more critical examination of forced retirement.

Thomas R. Klassen, with David Forgione (Chapter 4), extends the economic analysis in a new direction. Their consideration of the complexities of organized labour's positions on mandatory retirement makes a new and needed contribution to the public discussion. One of the accomplishments of the labour movement in the twentieth century was the establishment of retirement at a predetermined age with guarantees of a minimum level of income. While many unions continue to support this policy, especially large industrial unions, other elements of the labour movement have different views. Klassen and Forgione argue that post-industrial countries with weaker union movements seem to be eliminating compulsory retirement more quickly than Canada, and therefore the Canadian labour movement has an opportunity (and a responsibility) to help ensure greater flexibility in the evolution of the rules of retirement while protecting workers' interests and needs.

Allan Irving (Chapter 5) provides a fresh perspective on mandatory retirement, the only post-modernist analysis of which we are aware. The chapter analyzes our understanding of "old age" and retirement as social constructions that can be challenged and changed. Irving explores how the historical construction of categories affects the dominant contemporary images of aging. Using Foucault and others, he shows how our discourse about aging reveals our prejudices and suggests how these serve selected dominant interests.

Don Kerr and Roderic Beaujot (Chapter 6) offer the necessary demographic background to understand mandatory retirement. Demographics play a pivotal role in understanding our social life and organization,[25] and Kerr and Beaujot explain that the demographic processes used describing an aging population are more complex and more interesting than is widely understood. Because the Canadian demographic context of aging has changed considerably since the 1970s (for example, improvements in longevity and health and changes in labour force conditions), new social, employment, and retirement policies are needed. Kerr and Beaujot argue that it is important not to lose sight of the larger demographic context while debating the specifics of mandatory retirement, and they suggest that in the long run the changes to the Canadian age structure will have a stronger impact on the economy than adjustments to retirement age. Of particular note is their discussion of the relative importance of immigration. Canadian immigration is sometimes seen as either a source of major socio-economic problems or as the solution to such problems. Their analysis, however, argues that immigration is neither a threat nor a panacea, but rather has a relatively modest effect on population aging and the average age of the labour force.

Norene Pupo and Ann Duffy (Chapter 7) offer the broad social context in which to place the analysis of forced retirement by highlighting general employment trends. Canadian employees have diverse needs and interests, and this is at least as true for older workers as for younger ones. The increased pattern of participation in part-time employment, especially by women—and by extension many other marginalized workers—is part of the wider employment context for older and retired workers. Pupo and Duffy discuss how private and public lives intertwine and the interconnectedness of self-understanding, work, family, and home that are revealed in their research. Given the diversity of human lives and the complexities of the organization of human labour they conclude, not surprisingly, that both flexible retirement policies and flexible employment policies are needed. They also underscore the need for people to have economic security in order to fulfill many of their dreams.

Morley Gunderson and Douglas Hyatt (Chapter 8) argue for the value of the market-driven contractual arrangements inherent in many compulsory retirement schemes. Banning mandatory retirement implies prohibiting private contracting in this area, and they do not find the arguments for eliminating retirement at a fixed age to be persuasive enough to justify overriding contractual arrangements. They conclude with policy recommendations that provide for protection against age discrimination at all ages but exempt bona fide pension or retirement plans, thereby allowing for some degree of forced retirement. This chapter is particularly valuable because it defends some key aspects of mandatory retirement, and many of the other chapters, explicitly or implicitly, are responses to Gunderson and Hyatt's arguments.

Jonathan Kesselman (Chapter 9) directly challenges the Gunderson and Hyatt position. Kesselman argues that voluntary contractual agreements may be neither as voluntary as is assumed nor in the public interest. Moreover, greater flexibility in the organization of work can enhance the productivity of an older workforce, and human rights codes can be adjusted to accommodate individual private contracts where this is appropriate to allow necessary adjustments to disability insurance and pension benefits. In those jurisdictions where contractual mandatory retirement has been banned, it has been demonstrated that such a ban removes a key element of age discrimination without a large increase in employment of older workers. Just as Gunderson and Hyatt review and criticize the arguments against mandatory retirement, Kesselman considers and rejects the arguments made in favour of forced retirement.

John Munro (Chapter 10) examines the situation of Ontario universities generally, and especially that of the University of Toronto. He reviews and criticizes the main arguments for keeping contractual mandatory retirement, including those of the majority opinion in the 1990 *McKinney* case, which came out of Ontario. Munro contends that the arguments employed to support compulsory retirement in general do not withstand critical appraisal. This chapter presents a systematic refutation of these arguments in the context of a specific university. By virtue of its size, international status, and endowment, the University of Toronto is one of Canada's flagship universities. What does or doesn't happen there is important to all the other Canadian universities and to at least some other workplaces.

Jean McKenzie Leiper (Chapter 11) shares significant aspects of her biography, putting a human face and giving a personal voice to the analytic abstractions of many of the other chapters. Employment policies, labour practices, and court cases are ultimately about the meaning and significance of human experience. McKenzie Leiper illustrates well how compulsory retirement at an arbitrary age can be particularly disadvantageous to women. Her experience highlights that even existing institutional flexibility in contractual arrangements is not always, indeed not usually, exercised. Her narrative exemplifies the broader diversities of Canadian lives and suggests that legal structure and work arrangements need to be reconstructed by our own human imaginations and efforts.

In his conclusion, Peter Russell (Chapter 12) proposes that this book reflects the coming of age of a social movement that seeks social and legal reforms of ageist policies and practices in Canada. While mandatory retirement is the cutting edge of this movement, the true goal is nothing less than rethinking and restructuring the last third of our lives. He reminds us of the conditions for policy change, arguing that the courts follow the election results, and elected legislators respond to demographic and economic circumstances. The first step in the restructuring is the elimination of the legal stereotyping of older citizens, beginning with retiring mandatory retirement.

Overall, this volume expounds how ageism undermines the dignity of the person. Paid employment offers income security, meaning, and purpose for many people, some of whom are older workers. Forced retirement arbitrarily eliminates from paid employment precisely those who, by virtue of their longevity certainly, and often by virtue of their commitment and effort, have contributed the most to Canadian society.[26] Capricious termination of paid employment not only devastates individual lives; it also confounds at least the spirit, if not the letter, of the constitutional and human rights law that undergirds our national community. Arguing for the eradication of mandatory retirement is not a ground for lengthening the paid employment of most Canadians. Rather, it is an argument for justice for those who need or wish to work beyond a fixed age and for increased institutional adaptability for all employees to determine their own date of retirement. Flexibility in the organization of paid employment and retirement is necessary to meet the varied needs of an increasingly diverse population—to permit a variety of work arrangements in the later years of employment, to ensure adequate and secure income, to meet the labour needs of the economy, to enhance freedom for individuals, and to build and strengthen the Canadian community. Age sixty-five as a marker for forced retirement is arbitrary. Supporters and critics alike agree it is a form of age discrimination. The question is whether the practice can be justified.

2

The Ass and the Grasshopper: Canadian Universities and Mandatory Retirement

David MacGregor

An Ass, having heard some Grasshoppers chirping, was highly enchanted; and, desiring to possess the same charms of melody, demanded what sort of food they lived on to give them such beautiful voices. They replied, "The dew." The Ass resolved that he would live only upon dew, and in a short time died of hunger.[1]

For many, retirement signifies a sweet reward for decades of labour. Its attractions may explain why the average retirement age for Canadians has declined from sixty-five in the late 1970s to about sixty-one today. To a significant minority, however, retirement represents a poisoned chalice—displacement from primary sources of status, income, and meaning in life. This chapter focuses on the politics of eliminating mandatory retirement in universities, where faculty resistance has already upset—and may soon terminate—schemas of forced exit (see, for example, McKenzie Leiper, Chapter 11).[2]

University professors are ambivalent about retirement. The median age of retirement, about sixty-three years of age, is only slightly above that of the population as a whole. More than one-third of university teachers are over fifty-five years of age; fewer than 17 percent are under forty.[3] A growing minority of university teachers resents the absence of choice about when to leave. Simone de Beauvoir observed that ambivalence derives from the double-edged character of human labour:

> Work almost always has a double aspect: it is a bondage, a wearisome drudgery; but it is also a source of interest, a steadying element, a factor that helps integrate the worker with society. Retirement reflects this ambivalence, and it may be looked upon either as a prolonged holiday or as a rejection, a being thrown upon the scrap-heap.[4]

In Chapter 8, Gunderson and Hyatt compare forced retirement policies to student loans, mortgages, and marriage contracts. Marriage contracts, mortgages, and bank loans, however, may be suspended or renegotiated. Forced exit policies remove the choice to work or not—under equal conditions with younger peers—from those sixty-five and over. Negotiated compulsory retirement at many Canadian universities is likely a myth. Forced exit was never on the bargaining table at the University of Toronto—management unilaterally imposed it (Munro, Chapter 10). The same is true at the University of British Columbia, where the board of governors set the terms for retirement in 1963.[5]

Constriction of choice for older workers infects the language. In a landmark 1973 dissent, Supreme Court Justice Bora Laskin asked why unilateral termination of service on the basis of age should not be called dismissal merely because the employer refers to it as retirement. The verb retire is both transitive and intransitive, he observed. A worker may retire, or *be* retired. In the first case, retirement results from a decision by the employee; in the second, retirement is forced on the worker and equivalent to dismissal without just cause. In ordinary English, suggested Laskin, the worker is deprived of a job. The argot of retirement eludes this critical distinction.[6]

In the landmark *McKinney* and *Dickason* Supreme Court decisions, university administrators likely were proxies for investors and big employers who wished to preserve an entitlement to jettison older, more expensive employees. The court played to the neo-liberal mantra of cost reduction and control over labour costs.

Ageist power structures have a significant impact on labour market behaviour. A recent Statistics Canada study of 25,000 retired workers suggests as many as 60 percent would have preferred to keep working.[7] Forty percent mentioned mandatory retirement policies or corporate downsizing as the reasons they left work.

Fears that eliminating forced retirement will lead to a flood of less competent, elderly university professors are unfounded. Faculty working full time past sixty-five, where this is possible, are highly committed teachers and among the most productive researchers. Research in the United States indicates that elimination of mandatory retirement strengthened employers' long-term commitment to older workers. Abolition of forced exit in American universities produced none of the negative consequences predicted by supporters of compulsory termination policies.

Forced exit policies have renewed a practice that stirred universities during the hiring binge of the 1960s and 1970s.[8] With only a limited number of new Canadian PhD graduates available, university administrators have reverted to large-scale hiring of American and British faculty. In 2003, university professors formed the third-largest occupational group (1,560) among temporary workers admitted to Canada because of shortages.[9]

Mandatory Retirement, Universities, and Sociological Ambivalence

Ambivalence about retirement transcends gender, class, ethnicity, and occupational group. Eleven to fifteen percent of retiring Canadian workers may be affected by compulsory exit policies.[10] Almost one-third of non-retired Canadians aged forty-five to fifty-nine either do not know when they will retire or do not plan to retire.[11] Many retired Canadians may be displaced workers over the age of fifty-five who have given up looking for jobs. Retirement fluctuations in the 1990s are likely a result of government cuts and corporate downsizing.[12]

Ambivalence permeates Canadian business as well. Some employers view compulsory exit as a necessity; others are open to change.[13] Uncertainty is illustrated by the case of Thomas D'Aquino, who opposes compulsory retirement, while the organization he heads—the powerful Canadian Council of Chief Executives—is undecided.[14] The Conference Board of Canada supports eliminating forced retirement;[15] the Canadian Vehicle Manufacturers Association and the Toronto Board of Trade are opposed.[16]

Similar ferment roils political parties, voluntary organizations, and large sectors of the labour movement.[17] In Ontario, both the governing Liberals and the Progressive Conservatives (Tories) have declared against mandatory retirement. The New Democratic Party (NDP) applauds termination at age sixty-five, contending its removal would violate contractual agreements and weaken support for workers' pensions. The NDP position reprises that of large industrial unions, such as the Canadian Auto Workers and the United Steelworkers of America (see Klassen and Forgione, Chapter 4). These labour arguments are odd given that Quebec, the most heavily unionized province, abolished compulsory exit two decades ago without ill effects predicted by the NDP and its union allies.

Does Mandatory Retirement Favour Lower-Income Workers?

The labour perspective admits that compulsory exit harms women and minority groups, who may not have worked long enough to build a respectable nest egg. However, abandoning forced exit would tempt government to raise the age for retirement benefits (see Klassen, Chapter 4; Munro, Chapter 10). In addition, those resisting mandatory retirement are a small minority of mostly well-off workers, such as university teachers (see Gunderson and Hyatt, Chapter 8).

High-income occupations have no monopoly on resistance to compulsory exit. Affluent workers and those with rich pension plans are *least* likely to oppose mandatory retirement. Opposition to forced exit is more often found among less well-off workers and those with insufficient pensions.[18] A 2001 British Columbia Court of Appeal decision, throwing doubt on *McKinney* and *Dickason*, concerned a newly hired BC waste plant operator who was fired after his employer discovered that he was 65½ years old.[19] A Toronto firefighter, shut out at age sixty, recently brought his case to the Ontario Human Rights Commission.[20] The Canadian Media Guild is fighting the Canadian Broadcasting Corporation on behalf of Don Nelson, a former Anglican priest, who was terminated at age sixty-five from his $27,500 per annum position with the public broadcaster in early 2003.[21] Melrose Scott, a New Brunswick potash miner, wants to stay in his job after age sixty-five.[22]

The labour perspective underplays the devastating personal and psychological costs of mandatory retirement and forgets that a privileged retired worker dependent on a shrinking pension will eventually become a disadvantaged elderly person. After eliminating mandatory retirement in 1981, Quebec *lowered* the age of pension entitlement.[23] Changes in 1983 to the United States social security system that will lift the pension eligibility age from sixty-five to sixty-seven by the year 2027 were inspired not by elimination of mandatory retirement—banned in 1986—but by fears that baby-boomer retirees would strain government pensions. In contrast to Canadian organized labour, the British Trade Union Congress in September 2004 opted to fight compulsory exit.[24]

Gunderson and Hyatt contend that eliminating mandatory retirement would encourage employers to ditch older workers. Their argument is not supported by US data. Laws against age discrimination boost relative employment of older employees and allow them to stay longer.[25] Following elimination of forced exit in the United States, workforce participation of those over sixty-five rose significantly.[26] Elimination of forced exit in Quebec and Manitoba had a negligible effect on employment of older workers.[27]

Until the late 1970s, Canadian courts routinely supported collective agreements and company rules forcing women to retire earlier than men, typically at sixty instead of sixty-five. Women, it was held, clearly benefited from reduced years of work—they had more leisure, additional family time, and so forth. The Ontario Federation of Labour in 1973 concluded that women had a good thing going, and it was only a matter of obtaining the same socially progressive right for men.[28]

Retirement today—forced or not—is similarly cast as an opportunity to enjoy volunteering, recreation, and constructive leisure. Indeed, banks and other financial institutions saturate the media with advertisements promoting happy retirement outcomes. However, recent Statistics Canada reports question this image. For many Canadians retirement is largely unwanted, or spent in passive activities such as watching television or sleeping late.[29] Almost 60 percent of those who retired between 1992 and 2002 indicated they would have stayed at work under different circumstances.[30] These studies echo the findings of a large Canadian industrial firm that pioneered a survey of retirees in the 1970s. The company's medical research unit found many retirees were socially alienated and suffered rapid physical and mental deterioration after leaving their jobs.[31]

Universities Resist Change

New ideas about aging have barely penetrated university executive offices, where forced retirement is seen as the Rosetta stone of higher education, the key to managing a recalcitrant workforce. This calcified view resembles that held by university officials in the United States more than a decade ago, when, for example, Stanford President Gerhard Casper sonorously warned that "the end of mandatory retirement for university faculty was an ill-advised social policy that in the long run would hurt America's competitiveness."[32] In one of the most ironic symptoms, Simon Fraser University threatened to eject, on account of age, Professor Gloria Gutman, director of the university's Gerontology Research Centre and president of the International Association of Gerontology.[33]

University administrators provoke age and gender divisions[34] by pointing to unpleasant consequences if compulsory retirement is lifted. When the Ontario Tories introduced legislation to eliminate forced retirement in May 2003, the Council of Ontario Universities lobbied for a lengthy exemption. Universities launched a similar postponement effort following the new Liberal government's promise in early 2004 to end mandatory retirement. A suitable delay (of five years or more) would permit universities to offload most older professors.

In 1987, the Ontario government briefly considered banning compulsory retirement. Universities, following the successful example of their American counterparts,[35] appealed for a long transition period. The government initiative subsequently fizzled out. Universities sent their lawyers and labour policy experts to the Supreme Court in the 1990 *McKinney* and 1992 *Dickason* cases.[36] The Court famously ruled twice in favour of terminating professors at age sixty-five, stalling a national movement against age discrimination.[37] A last-ditch attempt to derail mandatory retirement in Ontario took the form of a private member's bill introduced in 1992. The bill died, but not before Dr. Arthur Kruger, head of the Ontario Institute for Studies in Education, warned against the dangers of ending forced retirement, using the image of a drunken professor sitting before his class with a thumb in his mouth.[38]

Universities across Canada (with the singular exception of Quebec) have adopted universal and strict enforcement of mandatory retirement.[39] Manitoba—which abolished forced retirement in 1982—is a paradigm case.[40] Angered by a bitter and successful faculty strike in the mid-1990s, "Tory education minister Linda McIntosh ... amended the acts governing the University of Manitoba, University of Winnipeg and Brandon University to allow compulsory retirement at sixty-five or older or through mutual agreement of the school and its employees." Premier Gary Doer left the legislation in place when his NDP government came to power in 1999.[41] Administrators in 2001 finally obtained the right, in collective bargaining, to enforce mandatory retirement at age sixty-nine at the University of Manitoba (with a provision for gradual retirement after age sixty-nine) and at the University of Winnipeg in 2002. The collective agreement at the University of Brandon does not include forced exit.

Manitoba's return to the practice is paralleled in Ontario, where higher education institutions that once had looser retirement rules (e.g., York University[42]) or did not enforce mandatory retirement (e.g., Carleton University, Ontario Institute for Studies in Education) have, over the past decade, imposed age caps. The University of Prince Edward Island adopted forced retirement in 1995. Alberta removed forced exit from the public service in the early 1980s, but allows it for universities. Although it has no compulsory retirement, the University of Calgary put the issue on the table in negotiations with faculty in

2002.[43] The English Canadian environment is reminiscent of the United States thirty years ago, when mandatory termination policies at universities were almost universal.[44]

Forced exit at age sixty-five is a recent phenomenon in Canadian higher education.[45] In 1959, for example, seventy-five-year-old Ivan C. Rand became the first dean of the newly founded law school at the University of Western Ontario. Rand had just retired from the Supreme Court.[46] Until the early 1970s, the retirement age at the University of Toronto was sixty-eight (see Munro, Chapter 10). The University of New Brunswick did not move to compulsory termination at sixty-five until around the same period; its Web site includes an ageist homily on forced exit.[47]

University teaching and research requires years of specialized education and careers often do not begin until the early thirties, or even later (see McKenzie Leiper, Chapter 11).[48] Pension plans for faculty, as for many Canadians, are frequently not indexed to the rate of inflation, so pensions may lose value quickly after retirement. Strikingly, the average age of retirement for university teachers (63) is lower than it is for other professional, scientific, and technical workers (age 64.6).[49]

Sociological Ambivalence and Social Change

Retirement is disguised as a form of individual choice when it is actually the result of social structural processes that manufacture ambivalence and reproduce inequality.[50] Forced disengagement from work imposes liabilities that escape notice by a younger employee because these losses reside in the future—"loss of identity, of power, of challenge to one's human abilities, of social ties and status, of simply being a part of the active mainstream."[51] As individuals grow older, circumstances and desires may develop in unexpected directions.[52]

Potential for social change increases "when substantial numbers of individuals who share a similar position attempt to negotiate the ambivalence created by current structural arrangements."[53] In Canada, formerly a leader in the struggle against age discrimination, potential for change may be reaching critical mass.[54] Based on force—the reluctant retiree is abruptly stripped of a job and its attendant rights and privileges—mandatory retirement also relies considerably on consent.

Erosion of consent and an attack on ageism are features of contemporary events in the universities and the larger society. A sign of this change may be favourable national media coverage given recently to eliminating forced retirement. Another is the unprecedented level of debate on the subject in Canadian universities such as the University of Toronto, the University of British Columbia, York University, Simon Fraser University, and the University of Alberta.

Re-evaluating Mandatory Retirement

Re-evaluation of forced exit policies in Canada forms part of an extraordinary and rapid transformation in thinking about population aging. Organisation for Economic Co-operation and Development (OECD), World Bank, and United Nations experts have replaced fear of a looming age crunch with a fresh perspective that the wave of retiring baby-boomers may bring major economic and social progress. Expanding the labour force participation of older workers, which fell steadily from the end of the Second World War to the mid-1990s, is a central feature of a new policy agenda.[55]

From 1975 to the late 1990s, governments in high-income countries—influenced by worry about a plague of old age[56] and high youth unemployment—colluded with business and unions to "concentrate the burden of unemployment on older workers."[57] Canadian universities encouraged early retirement and replaced only about 50 percent of departing faculty.[58]

Concern about an aging workforce is exacerbated by past Canadian recruitment patterns, particularly in the education and health sectors. Large numbers of workers hired in the late 1960s and 1970s are on the verge of retirement. Universities present a spectacular example. Twenty thousand professors will be replaced by 2010, and new faculty numbers will reach 30,000—close to the current total.[59]

Pressure to eliminate mandatory retirement comes from huge numbers of older people confronted by transformations in work and family life that have splintered rigid twentieth-century schemas of age/gender-related progression through school, work, and retirement.[60] Forced retirement depends radically on the profile of a typical retiree that blossomed between 1975 and 1995. The retiree was male, the main or only breadwinner, head of a household that no longer included children, mortgage-free, a believer in the golf-ball strewn ideology of the "golden years," likely to die relatively shortly, or already ill, and a member of a very small, comparatively powerless minority (those sixty-five and over).

Today, a sixty-four-year-old employee may be either male or female (one-third of employed seniors are women), may have a spouse with a separate career (and thus not available to accompany the retiree into retirement), may be the parent of children still requiring financial support, may have a large mortgage, may not believe in the ideology of the golden years, may love his/her job and enjoy a productive career, may be as healthy as someone in their mid-twenties and likely to live another 20–30 years, and belong to a growing and influential peer group (those over sixty-four, which grew 11 percent between 1996 and 2001).[61]

These factors contributed to a dramatic change in labour force participation rates of Canadian workers over age sixty-five, which, since 1996, have outstripped population growth, 20 percent compared to 11 percent. Older workers show up in a wider variety of occupations that reflect a diversified job market with broad requirements of education, experience, and skill.[62] Interestingly, truck driving posted the largest single increase for proportion of older workers, rising to the tenth most common occupation for those over sixty-five (See Pupo and Duffy, Chapter 7).

Ageism, the Supreme Court, and Structural Power Relations

Forced departure exemplifies ageism, which contains two potent and interconnected dimensions: a body of beliefs, stereotypes and attitudes; and discrimination, behaviour that excludes older people and treats them as lesser human beings relative to others on the basis of age.[63] These aspects of ageism feed on each other. Ageism oppresses older individuals by limiting opportunities and shaping how they are seen by people who embrace ageist stereotypes.[64] Similar to social class relations, notes Julie McMullin, forced retirement appears to be "a 'natural' result of a sifting process based on talent and ability or a more complicated process involving societal needs and differential socialization patterns." However, understanding mandatory retirement requires "an emphasis on structured forms of power, organization, direction, and regulation that exist in modern societies and through which ruling groups maintain and reproduce their dominant positions."[65]

Anna Freud said that forced retirement is easy to explain: it is the all-too-human desire for eminence, for succession[66]—a happy situation for aspiring labour kingpins and corporate moguls alike, whose older rivals will be sent to the ash heap at sixty-five. More insidiously, the powerful escape ageism (see McKenzie Leiper, Chapter 11). Prominent individuals, commented Peter Laslett, do not get old in the same way as the rest of us; they rarely experience the negative impact of ageism, and thus feel no compelling need to eliminate it from society.[67] We cannot discriminate against the old because we all become elderly, said the Supreme Court justices, hiding their smiles.[68] Justice Peter Cory, who wrote the *Dickason* decision at age sixty-eight, took over as Chancellor of York University in 2004, twelve years later, at age eighty.

Are Universities Unique?

The *McKinney* decision did not make a special case for universities but aimed to develop a judicial framework applicable to every workplace. *The Court dealt with sociological ambivalence by outlawing choice at age sixty-five.* The Supreme Court noted that labour organizations, such as the Canadian Labour Congress, were in favour of keeping forced retirement at sixty-five.[69] Although the 1992 *Dickason* ruling, like *McKinney*, specifically tested the application of mandatory retirement in the cloistered realm of the university, it also aimed at deciding the scope of mandatory retirement elsewhere.[70] Arguments in favour of forced exit common to universities are also used by other employers.[71]

The Supreme Court judgments drew a startling but consistent analogy between university teachers and factory workers, comparing tenure with workplace seniority systems. But the Court downplayed the cost savings that are a key reason for mandatory retirement policies. Thus—as McMullin and Marshall argue[72]—for employers of Montreal garment workers, *control over labour costs and workforce composition* is the chief benefit offered by forced exit and age discrimination policies generally. Better-paid, highly skilled older workers are jettisoned for cheaper, less qualified ones. By closing factories and then re-opening them under a new name, owners shed their older, unionized employees and replace them with younger, non-unionized contract workers. Jobless older workers encounter considerable difficulty in finding work, and if they succeed, their pay and benefits are much reduced.

Control over labour costs is equally a primary concern for university administrators. Universities hire adjunct teachers with no job security and poor pay to replace retired tenured faculty; sometimes the new adjunct is the formerly tenured professor. At the University of Manitoba, for instance, mandatory retirement was used to shed faculty and help make decisions on staffing.[73] A University of British Columbia Senate Report on the implications of removing mandatory retirement was unusually frank. It recommended rehiring professors terminated on account of age, provided this exacted no extra cost for the university.

> A candidate's age greater than 65 should not necessarily disqualify the individual from consideration for hiring if the individual would not cost the university any more than a new hire at the basic starting salary for the discipline. This would be particularly true in the absence of other qualified applicants for the position.[74]

Like elderly Montreal garment workers shifted into more vulnerable and lower-paid positions, faculty members over sixty-four are paid much less, lose job security, and are subjected to worse conditions than those enjoyed by younger tenured faculty.[75] Admitting that universities have a growing need to rehire retired faculty as adjuncts, University of Western Ontario president Paul Davenport cynically warned retirees they should "not … expect that luxuries like offices and parking will come with the post-retirement activities as 'difficult trade-offs' will have to be made to keep new full-time faculty."[76] An identical situation exists in the United Kingdom, where forcibly retired teachers are hired back by universities for less pay and with no employment rights.[77]

Whether garment workers or university teachers, victims of forced retirement are subject to irreversible and massive psychological and financial shock. "[M]any retired faculty suddenly discover that administrators treat them as if they'd undergone a form of 'absolute ontological change' the day they turned 65."[78] Later-life career disruption is significantly related to "poor self-rated health. This finding is consistent with the literature, which generally finds that unemployment is significantly associated with poor health."[79] Similarly, a British study of older people found compulsory exit from work has a pronounced negative impact on personal well-being. Individuals who exercised choice in the retirement decision were significantly healthier than those were were forced out.[80]

Management urged older garment workers who filed grievances to retire because they were "too old to work."[81] Ageist attitudes held alike by management and older employees themselves were harnessed to explain declines in productivity that actually resulted from owner-initiated changes in workplace organization. Employers convinced older workers to leave by introducing harsher working conditions. The value of older workers' experience was discounted by false assessments of productivity decline. McMullin and Marshall conclude that employers may use age to spark division and conflict in the working class, threatening solidarity among workers.

A similar process operates in the universities, where administrators indulge in ageist slogans and exploit divisions among faculty. A favoured administration tactic is to praise new faculty to the skies while celebrating the disposal of elderly scholars and teachers.[82] *McKinney* and *Dickason* relied heavily on demeaning imagery supplied by university administrators (see Munro, Chapter 10). Douglas Thorpe underlines the "clichés of modern anxiety about progress and competition" that appear in *McKinney* ("centres of excellence," "cutting edge," "faculty renewal"). The Supreme Court majority's "language and thought," reveal "fear that without mandatory retirement universities face degeneration and sterility."[83] Thorpe might have added that faculty renewal may be the consummate ageist term, implicitly comparing older faculty to a condemned neighbourhood in a rotted city core (see Irving's Chapter 5 on ageism, which describes the descent of sparkling cities into a dark chaos of aging and rot).

The title of an Association of Universities and Colleges of Canada (AUCC) report from March 2000 includes ageist stereotypes—"Revitalizing Universities through Faculty Renewal"—that are hammered home in the text.[84] Retirements allow deans to hire "the right type of faculty … who dovetail with their strategic priorities." Universities can improve "quality and diversity" by hiring "graduate students and young faculty." The AUCC gushes that "today's candidates for faculty appointment appear better equipped and qualified than ever before …. [M]ost deans and department chairs … were 'genuinely astonished at the superior capabilities of today's candidates for faculty appointments.'" For more than a decade, universities have experienced unacceptably high teaching loads, large class sizes and reduced administrative support for faculty but—as with garment workers, where productivity was harmed by poor management but blamed on older workers—the AUCC report does not mention that under these conditions, research and teaching of all faculty members, young and old, will be adversely affected.

Compulsory exit policies involve a process of estrangement that begins long before retirement, reducing efficient use of human resources, as the following testimony from a retired professor in British Columbia indicates:

> In my early sixties I was marvelling at American colleagues of my age, who were still contemplating career moves rather than retirement. It pointed up the absurdity of my situation, since I was fit (could and did run a half-marathon at 63), I was enjoying teaching, and also enjoying opening up a new seam of historical criticism. I haven't exhausted this seam yet, but archival work is necessary, and after six years of going to research libraries at my own expense without a salary it is near impossible to continue. As to retirement itself, I spent eight days in hospital in the summer of 2002 and two weeks in June 2003. I attribute both illnesses to the stresses of an unwanted retirement. In 2002, a lower-back injury left me helpless after a two-week spell of tackling our laurel hedges, a task I would not have undertaken but for loss of salary. In 2003 I suffered a life-threatening illness for

which no physical cause was found, but for which stress is known as a possible cause. Surely the known correlation between illness and unemployment holds also for unwanted retirement?[85]

There is scant evidence that young faculty outstrip the scholarly performance of older ones.[86] New professors cherish perspectives they learned from their mentors and examining committees, usually senior scholars with a breadth of experience.[87] This should not be surprising: most professions value "the accumulation of professional experience, expertise, and accomplishment" of older members. "It's a particularly tenacious misconception that younger faculty necessarily have fresher ideas than senior profs," says University of Alberta Graduate Student President Lee Skallerup. "Many are attracted to a department in the first place because of its strength in senior ranks." Dr. Cheryl Zusack, a new faculty member in the English Department, agrees. "As to whether or not senior faculty represent 'dead wood' in a department, I find this assumption rather odd given that in other professions the longer a person works in a field the more highly valued he or she becomes because of the accumulation of professional experience, expertise, and accomplishment."[88] Olive Dickason, the subject of the infamous *Dickason* case, entered postgraduate studies at age fifty and founded Canadian aboriginal history. Her case before the Supreme Court of Canada, which contested mandatory retirement at the University of Alberta, was lost on the single vote of a seventy-two-year-old justice.[89]

Tenure and Seniority

McKinney compared tenure with union seniority schemes. The Court opined that "[m]any organizations are so arranged that the individual is paid increasingly higher remuneration with the years with the expectation or understanding that he or she will depart at a certain stage."[90] The fatted goose must get the chop at sixty-five, regardless of occupation, or employment in the broader public or private sector. But a key US study disclosed no linkage between fixed termination and progressive pay scales in the universities.[91] Similarly, removal of compulsory retirement from other American workplaces did not weaken seniority systems. Instead, "the empirical evidence suggests that the predominant effect of age discrimination legislation has been to reduce the likelihood that firms renege on long-term commitments to older, higher-paid workers and consequently strengthen long-term relationships between workers and firms."[92]

In American universities, opinion has shifted from fear about older professors clogging the system to anxiety that with coming mass retirements, the corporate wisdom of older faculty will be lost. "What you lose is a generation deeply embedded in the culture of academic freedom and tenure ... the loss of that generation is the loss of institutional memory that many younger faculty have not had time to learn."[93]

Older workers are likely to stay in employment only a few years past the normal age of retirement. Even in university ivory towers, this is unlikely to disturb deferred-salary schemes and the gains in efficiency they are reputed to bring.[94] At Stanford University, for example, the average age of the professoriate was hardly touched by abolition of mandatory retirement in 1993.[95] The retirement age at the University of Calgary, which has no forced-exit policy, is only two years higher than at the University of Alberta.

American and Quebec universities (and other workplaces) have not installed expensive monitoring systems in reaction to eliminating mandatory retirement. The argument for increased monitoring assumes wrongly that older workers are less efficient. No sudden decline of abilities occurs at age sixty-five or above. Those with waning abilities due to health problems will most likely retire on their own. About 3,500 faculty took early retirement at Canadian universities between 1992 and 1997—some, no doubt, for health-related reasons.[96] The work life of older workers is relatively short, so that concern for productivity should rest on recent hires: they may be with the firm for many years and lack access to pensions that might encourage them to depart early.

Academic Freedom and Tenure

Universities insist that removing mandatory retirement would endanger academic freedom. It is a strange freedom purchased by sacrificing a minority's human rights. Forced retirement may be enough to strangle academic freedom in advance of retirement—as professors beg for favours the university might allow them in their powerless state following termination. "[I]f one of the purposes of universities is to produce a critical approach both to society and indeed to the universities themselves, surely it is those people who express that criticism who are least likely to receive the grace and favour of the administration in these regards, right?"[97] University teachers re-hired after age sixty-five, without regular employment rights, are especially vulnerable to attacks on academic freedom.[98]

In *McKinney*, the Court had substantial documentation from the United States, Quebec, and Manitoba to refute the notion that elimination of forced retirement would lead to loss of tenure. The majority chose to ignore the evidence in favour of an unexamined and crude tautology: mandatory retirement is necessary for tenure because mandatory retirement is necessary for tenure. "It is true," wrote Justice La Forest, "that many universities and colleges in the United States do not have a mandatory retirement but have maintained a tenure system. That does not affect the rationality of the policies, however, because mandatory retirement clearly supports the tenure system."[99] Large-scale studies in American universities reveal no trend against tenure. Faculty are staying in their jobs longer, rather than being subjected to attacks on tenure.[100] In the United Kingdom, where tenure was abolished more than a decade ago, enforced retirement remains in place. As in Canada, many tenured positions have disappeared in American universities, replaced by sessional appointments.[101] An important way to preserve tenure and academic freedom is to keep tenured professors, rather than forcing them out.[102]

Mandatory retirement and robust employment evaluation schemes have always existed together. According to United States Labor Department surveys in the 1980s, firms without compulsory retirement were also generally those with no institutionalized job evaluation.[103] Forced retirement and workplace evaluations are part and parcel of highly bureaucratized organizations, such as the modern university.

Obtaining tenure at a university may be one of the most arduous tests in the modern workplace. During the 1970s and 1980s, when funding and student numbers were in decline, Canadian universities turned away countless aspirants for tenure.[104] Few professors are fired for cause, but this is likely because stringent tenure proceedings weed out most of those not suited for academia. Tenured professors judged to be incompetent are often forced out by informal pressure from peers, department heads, and deans. Termination of faculty perceived to be at odds with workplace standards certainly occurs in universities, where "administrative mobbing" is not a rare occurrence.[105]

Faculty Renewal and Planning

McKinney and *Dickason* supposedly prepared the way for universities to replace retired faculty with younger scholars. Fresh from their hard-won court victory, however, administrators abolished positions of retiring faculty or replaced them with adjunct appointments. "Overall, Canadian universities have replaced only half of all departing faculty in the last few years."[106] Whether forced exit is retained or not, the large demand for faculty created by expected retirements and increased student enrolment will ensure that plenty of positions exist for the thin supply of youthful scholars.[107]

Mandatory retirement may be a job creation strategy for American and other foreign faculty rather than young Canadian scholars. In response to university lobbying, the federal government loosened its Canadians First policy, opening the gates to hundreds of foreign professors. "At UBC, about 40 percent of 188 hires in the 2002–03 academic year were foreign academics. In the 1996–97 academic year, it was 23 percent of 80 new hires."[108] Almost 50 percent of new hires at the University of Toronto in 2002–03 were foreign academics, compared to 31 percent in 1999–2000. Given the relatively small size of the Canadian academy, an "influx of a few hundred academics could significantly shift the balance of power in the university system."[109]

New ideas are most likely to emerge not through compulsory retirement but by encouraging faculty, at whatever age, to explore different paths. "Let's get beyond the ageism that is implicit in this kind of argument," says Lawrence Poston, Senior Associate Dean at the University of Illinois at Chicago. "We have all known young faculty members who are wed to older models of research and older faculty who have continued to be flexible and open to newer styles of inquiry."[110] Aged researchers, remarks Betty Friedan, continue to grow, and develop a stronger sense of morality—even at the risk of their careers or their life. The anti-nuclear movement was led by older scientists who came to abhor the atomic weapons research they did as young men.[111] Older researchers may be among the most original and productive faculty on campus. The major loss of productivity among faculty occurs in mid-career; no decline is registered in older age groups. Research performance may be stronger among faculty over sixty-five than for those ten or fifteen years younger. Teachers who enjoy inspiring their students "also tend to work to later ages, suggesting that strongly teaching-oriented faculty may also wish to continue working to advanced ages."[112]

Faculty older than seventy usually teach in the most elite private research universities where there is much pressure to publish and keep ahead in scholarship.[113] Indeed, Yale University was embarrassed into creating a centre for emeritus faculty after a former Yale professor, John B. Fenn, 85, who was forced to retire at 70, won the Nobel Prize for chemistry in 2002.[114] Yale's most prestigious molecular biophysicist, Dr. Robert G. Schulman, is 78. Dr. John Archibald Wheeler, 90, forced to retire from Princeton in the 1970s, partner of Einstein, and inventor of such terms as "black hole, quantum foam, law without law, to name a few" may be the most prominent and productive quantum theorist in the world.[115] The inventor of the laser, Charles H. Townes, is professor of astrophysics at University of California at Berkeley. Townes, 89, employs lasers to assemble images from faraway telescopes into a massive virtual lens.[116]

University administrators contend that mandatory retirement assists planning efforts—making it simpler to foresee shortages—and offers a convenient way to deal with deterioration caused by aging.[117] But early retirement schemes, much favoured by administrators, also disrupt planning. Moreover, in the new climate, where huge hiring gaps are expected, planning is likely to be thwarted not by uncertainty about retirement, but by competition for faculty from other universities and private sector employers. Faculty may depart unexpectedly because of illness or other causes. In any event, planning needs are an inadequate reason for forcing faculty into retirement.

University of Windsor human relations expert James Thacker notes that forced exit at sixty-five is a "lazy way" of planning an enterprise. Compulsory retirement is a blunt tool that casts out the best employees along with the worst. He advocates instead high-performance work systems, "conceptualized as a set of distinct but interrelated … practices that select, develop, retain and motivate the organization's workforce." Mandatory retirement regimes waste productive resources, lowering employee morale well before the retirement date, and casting out valuable workers.[118] The University of Toronto is reeling from the departure of high-profile teachers and researchers, attracted to American universities which have no mandatory retirement provisions.[119] Similar departures have hurt universities in British Columbia (Kesselman, Chapter 9). With nearly a third of faculty over fifty-five, the University of Alberta may have to drop its forced-exit policy in order to keep professors from bolting to institutions in the United States.[120]

American universities have reduced the ranks of senior professors with attractive retirement schemes or phased retirement programs.[121] About half of US universities have offered incentives since 1995 to encourage professors to retire before reaching seventy.[122] This is an especially effective approach, since "the most important predictors of retirement among faculty … were overall financial status and eligibility for full retirement benefits."[123] Almost 30 percent of higher education institutions employ phased-exit programs in which faculty work part-time before taking retirement.[124]

Lifting compulsory retirement in the United States had negligible results on the demand for new faculty. While many professors put off retirement, most still leave by their mid-sixties. Funding cuts accounted for much of the reduction in faculty during the last decade when tenured jobs were eliminated or converted into part-time positions.[125]

About 40 percent of faculty in private research institutions will remain until age seventy, "compared with about 25 percent at public research universities, and just under 25 percent at doctoral granting, comprehensive, and liberal arts institutions."[126] By 2002, only major research institutions had experienced significant growth in faculty over the age of seventy. Faculty over 70 make up 7 percent or less at most Ivy League American schools.[127]

Dorfman's study of "stayers and leavers" in an elite American university found a significantly higher proportion of women faculty worked past the retirement age. There was no difference in financial status between women who retired and those who stayed. "One possibility is that gender may have been interacting with family status. Although half of the employed women were currently married, none had any children or grandchildren, as compared to their retired counterparts, the majority of whom did." Regardless of gender, there is a higher proportion of married professors, and those with children and grandchildren, among retired than among employed professors. "It is not unreasonable to expect that older faculty who have spouses and perhaps far-flung children and grandchildren may want to spend more time with family members as they grow older. Such involvements may provide an alternative attraction to the academic enterprise as time becomes more finite."[128]

With large numbers approaching the normal age of retirement, resistance is growing. At University College of the Cariboo in Kamloops, BC, the faculty association filed a grievance against mandatory retirement.[129] A survey of academic staff at the University of Alberta revealed that most respondents felt that forced exit at age sixty-five is unfair.[130] The Simon Fraser University Faculty Association argues that compulsory retirement is an abuse of human rights. University of Toronto administrators and faculty have formed a working group on retirement policy.[131] Similarly, human rights cases on age discrimination involving university professors and other workers are coming forward in Ontario (See McKenzie Leiper, Chapter 11).[132]

Conclusion

As in other workplaces, mandatory retirement in the universities represents the socio-structural power relations of aging. Forced retirement cements the dominance of university administrators over faculty and also divides faculty members on age/gender lines, weakening their collective influence. These power relations are under siege and are likely to weaken as large numbers of faculty approach retirement age in the next few years. Prompt action by government to terminate forced-exit policies will relieve uncertainty and prevent a possible breakdown of relations between groups on either side of the mandatory retirement debate.

Exemptions to the Age Discrimination in Employment Act provided to American universities in 1978 and 1986 (see Chapter 10) occurred when few professors were at the normal retirement age. Moreover, the 1986 exemption for professors came after the retirement age had already been lifted to age seventy. The situation is strikingly different in Canada today, where large numbers of university teachers (perhaps as many as 20,000) are on the verge of forced retirement at age sixty-five. In Ontario alone, close to 400 faculty will reach retirement age annually over the next five to seven years.[133] Lifting mandatory retirement would partially relieve an expected staffing crisis.[134]

Fears that removing an age cap on employment would engulf universities in a flood of older faculty are unfounded. The US experience suggests that removing the age barrier from sixty-five increased the average retirement age only by a few years. Faculty rarely work past the age of seventy, and those who do remain employed beyond that age are among the best teachers and most formidable researchers. In Canada, among 800 faculty at Concordia University, where forced exit was eliminated twenty years ago, only a dozen or so are older than sixty-five; two are over seventy. Figures are similar at other Quebec universities.[135]

Lorraine Dorfman's study of the impact of lifting compulsory retirement illustrates the ass and the grasshopper fable chosen as a theme for this chapter. Even with the end of compulsory retirement in Canada, most university teachers will retire well before they reach age seventy. Retired and employed faculty are equally happy, but for very different reasons.

> [E]mployed faculty seemed to view their work as a "calling" which provided them with considerable pleasure. Retired members, on the other hand, were more ready to move on. They wanted time to do other things, were sometimes tired of their work, felt it was time to retire, or found changes in their departments or institution aversive.

Happiness among retired and employed professors "reflects at least in part satisfaction with the work and retirement choices they made."[136]

The costs of keeping older faculty for a few years are negligible compared to the benefits. Replacing a professor with a younger recruit yields only a small, short-term reduction in salary (about one-third at most). However, this gain must be balanced against the loss of the older professor's institutional knowledge and experience, and the corresponding need to train the inexperienced new teacher (See Chapter 10). "Knowledge transfer," asserts Graham Lowe, "may be one of the most powerful incentives for employers to take a more flexible approach [to retirement]."[137] Older teachers are more loyal than younger ones, who may depart for greener pastures soon after being hired, especially in a tight job market. As universities are depleted of experienced teachers due to retirement, mentorship from older faculty is urgently needed; otherwise, institutional memory will be lost.

Some administrators recognize the inevitability and benefits of change. In a speech in May 2003, the president of Dalhousie University, Tom Traves, said universities "may have to face … questions about the normal retirement age, [and the] rights and entitlements of retired professors."[138] Similarly, the University of British Columbia is struggling to come to terms with faculty shortages and the problem of compulsory retirement.[139] Allowing faculty to continue working past the normal age of retirement would help universities avoid faculty shortages and ensure that students are taught by experienced and qualified teachers.

A remarkable development regarding the issue of mandatory retirement in the universities is the December 2003 report of Quebec's Superior Council on Education.[140] The Council does not subscribe to fears expressed by university administrators outside of Quebec about an imminent shortfall in supply of faculty. But this may be partly because Quebec has eliminated mandatory retirement. The Council warns against a hiring spree such as occurred in the 1960s and 1970s, when many less-qualified faculty received teaching posts. Repetition of this error would create similar conditions in the future. A key recommendation is "to aim at a diversification of age groups at the time of hiring [alternation between beginning professors and experienced professors] in order to establish a continual process of renewal of the academic body in the future ...." This effort would include "various means ... to support the integration of newly recruited professors and to keep in place experienced professors. Accent could be put on certain measures such as ... mentorship of experienced professors towards recruits."

Universities must reflect the needs of the larger community, as recognized by efforts at diversity in post-secondary institutions. Diversity is offered as an excuse to keep compulsory retirement—forcibly retired faculty are replaced by younger teachers who are more likely to be female and representative of different ethnic groups. But diversity must include representation of all age groups, including those over sixty-four. Compulsory retirement of some of the country's most highly educated and valuable workers—who could provide a voice for an older generation—compromises diversity initiatives that must include age. Universities ought to join government in restoring the link between older people and civil society.

3

The Shifting Judicial Foundation of Legalized Age Discrimination

C.T. (Terry) Gillin & Thomas R. Klassen

Constitutional law embodies the moral aspirations of a community. The central objective of the Canadian Charter of Rights and Freedoms is the creation of a certain kind of political community for Canadians. The Charter seeks to protect individual rights and freedoms by proposing a national community of collective rights and freedoms.[1] Rights, then, are not essences of individuals; they are characteristics of human relations. Constitutional rights are principles defining how we should relate to one another. The Charter articulates fundamental law that guides Canadians in addressing the question, "What kind of community shall we be?"[2] In order to build the community envisioned by the Charter, the public, as well as scholars, need to engage its ideas in contemporary contexts. This chapter is one such effort; it focuses on the rights of workers at age sixty-five and primarily analyzes a series of Supreme Court of Canada judgments on mandatory retirement.

Canadian society has complex sources for its organization of retirement policies. Mandatory retirement policies generally reflect employment practices embodied in pension plans and collective agreements, within the context of human rights or other legislation, overseen by the judicial system. Business and labour policies and practices, federal and provincial legislation, and judicial determinations all influence the normal retirement age in a specific workplace or industry. In Canada, very few laws specify the retirement age for workers—with some exceptions, such as for airline pilots, military personnel, and judges. Unlike most nations, even those with federal systems of government, considerable responsibility for Canadian labour market regulation rests with the provinces. The judiciary interprets "property and civil rights," which are under provincial jurisdiction, as covering a very wide field including employment-related relationships. Approximately 90 percent of Canadian workers are under the jurisdiction of provincial legislation.

Mandatory retirement legal cases take two different forms: Workers forced to retire before age sixty-five, typically because of bona fide occupational requirements (BFOR), and those required to retire upon reaching age sixty-five. Human rights codes cover both situations, but this legislation varies considerably from jurisdiction to jurisdiction. For example, employers covered under the Canadian Human Rights Act can legally force workers to retire when they reach "the normal age of retirement" for employees performing the same type of work.[3] In Ontario, British Columbia, and several other provinces, it is legal to retire workers (who are employees of provincially regulated employers) at age sixty-five. Quebec and Manitoba prohibit mandatory retirement at any age, while in New Brunswick, termination of employment is linked to pension plan provisions. (The appendix to this chapter provides a summary of the current legislative provisions on mandatory retirement age for the federal, provincial, or territorial jurisdictions.) Mandatory retirement does not apply to Canadians who are self-employed and therefore do not typically face employer-mandated or legal restrictions on employment.

In the past quarter century, significant judicial activity focused on mandatory retirement. In the early 1980s, the Supreme Court of Canada seemed to assert a position that was skeptical and critical of forced retirement. The introduction of the Charter of Rights and Freedoms, whose equality provisions (section 15) took effect in 1985, ironically shifted judicial focus from broad human rights principles to a socio-economic analysis. Specifically, in 1990, the Supreme Court ruled that mandatory retirement at age sixty-five was constitutionally acceptable discrimination. More recently, particularly in some 1999 opinions, another shift seems to have occurred in judicial thinking on the principles applicable to the question of mandatory retirement.

Virtually all the articles in this book—and all informed discussion of the issue—consider and respond to the landmark 1990 *McKinney* case.[4] That this is so points to the need for careful analysis of the judicial and legal context generally. In this chapter, we proceed in three steps. First, we briefly consider the pre-Charter decisions by the Supreme Court to analyze the grounds of judicial determination. During this time, the Court established a legal position critical of involuntary retirement. Second, we trace the shift that occurred with *McKinney* and related cases in which the Court ruled that the practice of mandatory retirement at age sixty-five did not violate the Charter. This shift reflects the ascendancy of pragmatic socio-economic considerations and, possibly, ageist stereotyping. The beneficiaries of this judicial pragmatism seem to be employers. Third, we examine the developments of the past few years at the judiciary and in related public policy bodies concerning mandatory retirement. The more recent judicial opinion seems to re-emphasize the importance of human rights principles, but it is yet unclear whether this approach will be determinative of future retirement cases. We argue the future rests more with ordinary Canadians than with the courts.

Our analysis encompasses both the Charter and human rights legislation. The notable difference between these is that the Charter is entrenched constitutional law, whereas human rights codes are legislated, federally or provincially. The former protects basic individual and group rights and freedoms from inappropriate exercise of government power; the latter protect similar rights and freedoms in the wider civil society.

The broad meaning of the judicial opinions considered here is their version of the Canadian community they invite us to fashion.[5] Our analysis illustrates that the law is a social construction, and specifically highlights that the workplace arrangements constructed and supported by jurisprudence can be designed differently.[6] To the extent that recent court decisions are framed in terms of human rights, courts seem to be creating an opportunity for a new (or perhaps renewed) understanding of the rights of workers aged sixty-five and older. The role of analysts is to tease out the possibilities embedded in legal texts so that the everyday experience of rights and freedoms can be more fully realized.[7] We hope to encourage an interpretation that furthers the human rights of older workers in Canadian society.

Pre-Charter Supreme Court Decisions

Before the existence of the Canadian Charter of Rights and Freedoms, emergent federal and provincial human rights legislation, employment laws, common law, and judicial findings governed mandatory retirement. Comprehensive human rights legislation was initially instituted in the 1960s; for example, the Canadian Bill of Rights was passed in 1960 and the Ontario Human Rights Code in 1961, largely in response to the civil rights movements that challenged the stigmatizing effects of race and ethnicity. Legislation has since been introduced or amended in various jurisdictions to include gender, sexual orientation, and other putatively objective conditions that limit opportunities and freedoms. The focus on rights and freedoms derives from notions of natural justice and is articulated in constitutions and charters or bills of rights. The legislative result is a number of grounds—race, age, gender, disability, and others—on which discrimination, or the withholding of opportunities available to other workers or applicants for employment, was prohibited.

The leading Supreme Court case of the 1980s on the issue of mandatory retirement before the age of sixty-five focused on the question of bona fide occupational requirements (BFOR). In *Ontario Human Rights Commission v. Borough of Etobicoke (1982)*,[8] the Supreme Court staked out a critical position on ageism and mandatory retirement. The case addressed the legality of mandatory retirement at age sixty for Ontario firefighters. Under the Ontario human rights legislation, workplace discrimination based on age was illegal, with the protection having an upper limit of age sixty-five for the purposes of employment. On behalf of a unanimous Court, Justice McIntyre writes:[9]

> We all age chronologically at the same rate, but aging in what has been termed the functional sense proceeds at widely varying rates and is largely unpredictable. In cases where concern for the employee's capacity is largely economic … and the circumstances of employment require no special skills that may diminish significantly with aging, or involve any unusual dangers to employees or the public that may be compounded by aging, it may be difficult, if not impossible, to demonstrate that a mandatory retirement at a fixed age … may be validly imposed …

Keeping in mind the qualifying language of circumstances and risks, in this case the Court rejected ageist stereotypes and suggested that individual capacity to work, or the demonstrated limits of that capacity, must be the basis of judicial decisions.

The Court also provided a standard for testing the legality of when a BFOR can be used to justify forced retirement. The retirement rule or practice must meet both a subjective and an objective test. The subjective element is that the BFOR "must be imposed honestly, in good faith, and in the sincerely held belief that such limitation is imposed in the interests of the adequate performance of the work," and the objective part is that the requirement must be "reasonably necessary to assure efficient and economical performance."[10] In *Etobicoke*, the court determined that the employer had acted in good faith in the establishment of its age standard, but it could not meet the objective test because of the lack of scientific evidence. Moreover, the Court also argued that Ontario's human rights code constituted public policy, and as such, its provisions "may not be waived or varied by private contract."[11] In effect, the Court was recognizing human rights legislation as fundamental law that should not be relinquished even by contract arrived at through collective bargaining.

This decision proposed three key legal principles: (1) that it may be impossible to defend mandatory retirement on largely economic grounds; (2) that BFORs must meet both a subjective and an objective test; and (3) that human rights are fundamental law and cannot be contractually dispensed with. Only eight years later, in the case of *McKinney*, the Supreme Court would back away from these positions.

But before it changed direction, the Court reaffirmed the fundamental importance of human rights laws in another case.

In *Re Winnipeg School Division No. 1 v. Craton (1985)*,[12] the Supreme Court considered a conflict between Manitoba's Human Rights Act (1974), which prohibited age discrimination, and the Public Schools Act (1980), which empowered school boards to establish a compulsory retirement age. A teacher won a declaration from the trial court that the human rights legislation took precedence over the Schools Act, and therefore the collective agreement requiring mandatory retirement at age sixty-five contravened the Human Rights Act. The trial court's decision was affirmed on appeal by the Manitoba Court of Appeal, and reaffirmed by the Supreme Court of Canada.

In *Re Winnipeg*, the Supreme Court once again defended the special nature of human rights legislation (p. 6) and continued to express a general antipathy toward mandatory retirement, first articulated in *Etobicoke*. The Court confirmed a legal presumption against forced retirement, with the BFOR exception available under specific conditions. In the late 1980s, as other mandatory retirement cases made their way through the appellate courts of Ontario, British Columbia, and Alberta, it appeared likely to some that the Supreme Court would eliminate compulsory retirement altogether under the newly enacted Charter.[13]

Leading Decisions Under the Charter

The Charter was enacted in 1982 and the equality rights of section 15 came into effect in 1985. The main result was the substantial strengthening of individual rights in Canada. In the early 1980s some anticipated that, building on the earlier decisions emphasizing human rights, the new constitutional regime would view forced retirement at an arbitrary age as counter to the equality principles enunciated in the Charter.[14]

The interpretation of fundamental law expressed in the Charter can be considered the ultimate ground of determination for whether mandatory retirement is acceptable in Canadian society. The basic structure of the Charter is to provide fundamental and sweeping protections within a context of reasonable limits. Section 15(1) of the Charter:

> Every individual is equal before and under the law and has the right to the equal protection and equal benefit of the law without discrimination and, in particular, without discrimination based on race, national or ethnic origin, colour, religion, sex, age, or mental or physical disability.

These rights are exercised within the "reasonable limits" established by section 1:

> The Canadian Charter of Rights and Freedoms guarantees the rights and freedoms set out in it subject only to such reasonable limits prescribed by law as can be demonstrably justified in a free and democratic society.

A number of cases designed to test the application of the Charter reached the Supreme Court during the late 1980s.[15]

In 1990, the Supreme Court issued its first decisions vis-à-vis mandatory retirement: *McKinney v. University of Guelph*, *Harrison v. University of British Columbia*, and *Stoffman v. Vancouver General Hospital*.[16] The first two dealt with university professors and librarians, the last with physicians. In all three cases, the Court was asked to determine whether the Charter was applicable to mandatory retirement policies, i.e., whether such policies violated section 15(1) of the Charter's "equal protection and equal benefit" rights. Of the three, *McKinney* is the lead decision and will be the focus of our analysis.

In 1987, the case of seven Ontario professors and one university librarian who were facing forced retirement because they had reached the age of sixty-five was heard by the Ontario Court of Appeal, *Re McKinney and Board of Governors of the University of Guelph and eight other applications*.[17] The Ontario Court of Appeal had held that the Charter did not apply to universities because they were not part of government. The court further found that the provincial human rights code, under which age protection for purposes of employment expires at age sixty-five, did not offend the Charter.

When *McKinney* was appealed to the Supreme Court of Canada, the justices agreed that the differential treatment permitted by the restricted definition of age does offend the equality rights guaranteed in the Charter. Thus, at its core, this case is about discrimination.[18] The majority in *McKinney* reasoned that discrimination based on the enumerated grounds specified in the Charter is a social evil, and mandatory retirement deprives employees of a benefit under the Code based on age, a ground explicitly identified in the Charter. The majority determined, however, that the Charter could not rectify all evils. "Only government requires to be constitutionally shackled to preserve the rights of the individual," Justice La Forest wrote on behalf of the majority.[19] The Charter protects individuals from governmental abuse of power, not from the exercise of private power. Potential abuses of private power may be regulated by governmental agencies, such as human rights commissions, to prevent, limit, and control private offences against the rights of individuals. In short, since the universities were not government for purposes of the Charter, its provisions did not apply to them, and relying on section 15(1) could not eliminate involuntary retirement at age sixty-five.

The Court found that the universities did not have an institutional link to government and, therefore, the Charter did not apply; nevertheless, the Court considered the issue of mandatory retirement socially pressing and furthered its analysis by examining the constitutional status of the human rights legislation. In *McKinney*, the Court focused not only on the equality rights guaranteed in the Charter but also on whether Ontario's human rights code, which permits employment discrimination against employees at age sixty-five, meets the test of constitutional validity contained in the Charter. In Ontario, section 4(1) of the Human Rights Code, 1981, provided for equal treatment:

> Every person has a right to equal treatment with respect to employment without discrimination because of race, ancestry, place of origin, colour, ethnic origin, citizenship, creed, sex, sexual orientation, age, record of offences, marital status, same-sex partnership status, family status or disability.

For the purposes of employment, however, the Ontario code specified "age" means an age that is eighteen years or more, except in subsection 4(1) where "age" means an age that is eighteen years or more and less than sixty-five years" (s. 9[a]). The result is that employees can be required to retire at age sixty-five *and* those employees who continue to work after age sixty-five do not have to receive equal treatment and opportunities in such matters as work conditions, workload, salary, and benefits.

Although the Supreme Court determined that the Ontario code had the effect of permitting forced retirement and did discriminate contrary to section 15(1) of the Charter, such a policy was saved by section 1 as a "reasonable limit." Though differential treatment based on age is an affront to the principle of equality, the Court rationalized the discrimination with a socio-economic argument. The majority of the Court argued that retirement is a by-product of modern society, that "65 has now become generally accepted as the 'normal' age of retirement," and that "mandatory retirement has become part of the very fabric of the organization of the labour market in this country."[20] Moreover, age is to be distinguished from the other enumerated grounds. Justice La Forest writes, "There is a general relationship between advancing age and declining ability," whereas no such correlation exists between any of the other specified grounds (e.g., race, colour, or religion) and ability.[21] Justice La Forest argued that in allocating social benefits, it is acceptable to balance the competing needs of different generations.

Justice La Forest established that the context for the discussion of retirement policy is the importance of work in society.[22] He cited *Ref. Re Public Service Employee Relations Act* (1987),[23]

> Work is one of the most fundamental aspects in a person's life, providing the individual with a means of financial support and, as importantly, a contributory role in society. A person's employment is an essential component of his or her sense of identity, self-worth and emotional well-being.

Given its fundamental role in human life, work should not be taken away without significant justification. Justice La Forest's view, however, was that compulsory retirement at age sixty-five is justified because it has become part of the usual organization of labour. Based on the

Ontario Appeal Court's arguments in *McKinney*, the Supreme Court of Canada saw mandatory retirement as allowing deferred compensation, facilitating recruitment, and avoiding continuous productivity reviews, while permitting both employers and employees to plan for their financial futures. The key legal elements of *McKinney* are summarized in Table 1.

Table 1: Summary of the Opinions in *McKinney* (1990)

	OPINIONS			
	La Forest, Dickson, Gonthier and Sopinka	*Cory*	*Wilson*	*L'Heureux-Dubé*
Does Charter apply to retirement provisions of universities?	no	yes	yes	no
If yes, does mandatory retirement violate s. 15 of Charter?	yes	yes	yes	no need to answer
Is the violation justifiable under s. 1 of Charter?	yes	yes	no	no need to answer
Does s. 9 of the Ontario Human Rights Code violate s. 15 of Charter?	yes	yes	yes	yes
Is the violation justified under s. 1 of Charter?	yes	yes	no	no

Section 9 of the Ontario Human Rights Code protected against discrimination in employment of persons ages eighteen and above but less than sixty-five.

The dissenting voices, both female Justices on the Court, took issue with key points in the majority opinion. Justice Bertha Wilson reasoned that section 9(a) of the Ontario Human Rights Code was not saved by section 1 of the Charter. She argued that because the code fails to distinguish between those who are able to work and those who are not, a rational connection between the policy and its objectives had not been adequately established. She writes that section 9(a) "operates to perpetuate the stereotype of older persons as unproductive, inefficient, and lacking in competence." Moreover, Justice Wilson remained unpersuaded by her majority colleagues that free collective bargaining justified the discrimination, noting "the vast majority of the workforce is unorganized."[24] She remarked further that immigrants, women, and the unskilled—"the most vulnerable employees"—would be the most affected by the lack of legislative protection.[25]

Similarly, Justice Claire L'Heureux-Dubé found that section 9(a) of Ontario's code was not saved by section 1 of the Charter. She argued that the statute's restricted age definition "is inconsistent with the fundamental values enshrined in s. 15(1), the protection and enhancement of human dignity, the promotion of equal opportunity, and the development of human potential based upon individual ability."[26] In the absence of any evidence concerning an individual's ability, age discrimination is unwarranted. The reach of section 9(a) is overbroad, prohibiting not only complaints of forced retirement but of any form of employment discrimination, including hiring, demotion, transfer, or salary reduction. In a twist on Justice La Forest's emphasis on the importance of work, Justice L'Heureux-Dubé writes:[27]

> [I]f 'in a work-oriented society, work is inextricably tied to the individual's self-identity and self-worth', does this mean that upon reaching sixty-five a person's interest in self-identity and stake in self-worth disappear? That is precisely when these values become most crucial, and when individuals become particularly vulnerable to perceived diminutions in their ability to contribute to society."

Two years later, in 1992, the analysis provided by the majority in *McKinney* continued to dominate the Court's reasoning about mandatory retirement. In *Dickason v. Governors of the University of Alberta et al.* (1992),[28] a faculty member challenged involuntary retirement under Alberta's human rights legislation. The parties to the dispute agreed that the policy of compulsory retirement contravened the provincial act; the question was whether the practice was justified as a BFOR.

In *Dickason*, the majority of the Court again applied the arguments developed in *McKinney*.[29] Justice Cory authored the majority opinion and argued that a collective agreement that includes mandatory retirement at a fixed age provides evidence of the reasonableness of compulsory retirement, especially in cases where it is linked to tenure. Once again, age is conceived as distinguishable from other grounds of discrimination, reducing the level of defence needed to justify forced retirement. As in *McKinney*, the socio-economic objectives of the retirement policy are seen as sufficient to accept the limitation on the Charter's guarantee of equality.[30] The Court was not unanimous, with dissents from Justices L'Heureux-Dubé, McLachlin, and Sopinka—the last was part of the majority in *McKinney*. Overall, the Court confirmed its shift on the issue of mandatory retirement from a broad human rights approach to a narrow socio-economic interpretation.[31]

The *Dickason* case is notable in part for its confirmation of the lower standard established by the Supreme Court to defend against age discrimination. An analysis of whether a challenged law is acceptable as a "reasonable limit" under section 1 of the Charter relied upon principles first articulated in *R. v. Oakes* (1986),[32] an early Charter case.[33] *Oakes* established that the onus rests with the party seeking to uphold the limitation of a Charter right. The objectives of the law in question must be sufficiently important to justify the limitation of constitutional rights, and the measures specified in the law must be proportional to the objectives sought. The proportionality test has three aspects: (1) The limiting measures must be rationally connected to the objectives; (2) they must minimally impair the Charter right, (3) and the effects of the limitation must not be so severe that the abridgement of rights outweighs the objectives sought. In sum, the effects of the legal medicine must not be worse than the social problem that needs healing.

In *Irwin Toy Ltd. v. Québec (Attorney General)* (1989),[34] the Supreme

Court determined that different degrees of probable impairment of rights were possible. Minimal impairment will "vary depending on the government objective and on the means available to achieve it."[35] This approach permitted the more relaxed test for impairment that was used in both *McKinney* and *Dickason*. In both cases, the majority of the Court found that the government had a reasonable basis for believing that mandatory retirement only minimally impaired the rights of employees. The reasonable basis was established because—excellence being vital to universities—faculty renewal is required if universities are to remain on the cutting edge of new knowledge. Mandatory retirement facilitates faculty renewal. Moreover, academic freedom is equally essential, and mandatory retirement provides for a significant measure of freedom with a minimum of performance reviews. The critical issue here was whether the relaxed standard is appropriate. Justice Wilson, in dissent, argued that if one of the purposes was to provide opportunities for a vulnerable group, such as visible minorities, that would provide a legitimate foundation for a relaxed standard, but young academics are not a vulnerable group.

The Court's attitude toward ageist stereotypes is troubling in these decisions. Justice L'Heureux-Dubé, in her dissent in *Dickason* (1992), attacked the ageism inherent in the majority opinion:[36]

> Because, in our society, old age tends to be less associated with wisdom and tranquillity and more with infirmity and dependence, we fear it. We may be more likely to discriminate against elderly people, in a futile attempt to distance ourselves from what will inevitably occur to each one of us.

She went on to observe that, while human rights legislation prohibits the use of stereotypes associated with a number of ascribed characteristics, "one of the most persistent prejudices in our society is that old people are uniformly senile and physically incapacitated."[37] Fundamental human rights are at stake, not just the socio-economic implications of working beyond age sixty-five.[38] The Supreme Court's acceptance of the stereotype that loss of ability is directly correlated with age is radically inconsistent with human dignity, which is at the core of the Charter's provision for "equal protection and equal benefit of the law."

The most recent Supreme Court of Canada decision directly addressing mandatory retirement is, at the time of this writing, almost a decade old. In *Re Large et al. and the City of Stratford et al.* (1995),[39] the Court returned to questions regarding bona fide occupational requirements. A police officer in Stratford, Ontario, faced mandatory retirement at age sixty—a provision in his collective agreement. The Ontario Human Rights Commission found the provision of the collective agreement contravened the Human Rights Code. The Divisional Court and the Court of Appeal upheld the findings of the Commission's board of inquiry. However, the Supreme Court overturned the decisions made below it and allowed the City of Stratford's defence of its mandatory retirement policy based substantially on the fact that the policy was a provision of the collective agreement and was union-driven.[40] Moreover, the Court rejected the idea that individual accommodation was necessary, conceiving it as an "impermissible extension" of legal principles set out in earlier cases.[41]

It is striking that the *Large* ruling contradicts the result in *Etobicoke*. In the 1982 *Etobicoke* decision, before the Charter took effect, the Supreme Court emphasized the human rights legislation was fundamental law and that clear proof of danger to public safety was required to establish a BFOR. In 1995, under the aegis of the Charter, the *Large* decision emphasized that mandatory retirement policy was justified as a provision of a collective agreement and that the BFOR does not require individual accommodation. The presumption had shifted from protecting the individual right to work, and the dignity associated with employment, to assuming contractual arrangements justify at least some types of discrimination. The underlying legal implication was that just because it is possible for an employer to avoid a discriminatory practice, it does not mean that it is necessary to do so.[42]

In *Large*, the majority of the Court held that an employer could meet the subjective "good faith" element of the BFOR test without "belief in the *necessity* of the work-related requirement," especially when the rule was a product of collective negotiation.[43] Moreover, the objective element requiring that the rule be "reasonably necessary" could be met without requiring the employer to adjust job duties or individually test employees.[44] The Court's reasoning in *Large* weakened the standard originally articulated in *Etobicoke*. The rule of necessity became merely a rule of relevance. Under *Large,* simply demonstrating some relevance between the policy and the purposes of the enterprise was sufficient; the employer was not held to a higher standard of demonstrating either the necessity of the policy or the difficulties in designing a less burdensome policy, such as adjusting job duties or testing individuals.[45] The legal analysis in *Large* was generally consistent with the pragmatic socio-economic analysis found in *McKinney* and companion cases, disregarded the earlier emphasis on protecting human rights, and confirmed the general trend for the Supreme Court to leave unchanged mandatory retirement regulations and legislation. *McKinney* has received a great deal of analytic attention; but by virtue of its limited application and the shifting grounds of subsequent jurisprudence, it may be an outmoded standard.

Subsequent Jurisprudence

Subsequent mandatory retirement cases focus on different aspects of the judicial analysis: first, the extent to which the *McKinney* decision settled the question of the constitutionality of mandatory retirement; second, the extent to which BFORS continue to be important, and third, the extent to which judicial law has become more unified in its approach to involuntary retirement at age sixty-five. Although *McKinney* and its companions remain the Supreme Court's lead rulings on mandatory retirement, they are, like all judicial decisions, highly circumscribed. The decisions apply only to universities in Ontario and Alberta, and community colleges and hospitals in British Columbia. These decisions, although often interpreted as sanctioning compulsory retirement, in fact do not comprise a general rule that mandatory retirement policies can always be justified under section 1 of the Charter. *Tétreault-Gadoury v. Canada (Employment and Immigration Commission)* [1991][46] highlights the importance of case-by-case consideration. Marcelle Tétreault-Gadoury lost her job and was denied ordinary unemployment insurance benefits on the basis that she was over sixty-five. Justice La Forest, writing on behalf of a Court united in its conclusion, found that section 31 of the

Unemployment Insurance Act of 1971 violated the Charter in terminating benefits at age sixty-five. Justice La Forest wrote:[47]

> The most harmful and singular aspect of section 31 of the Act is that it permanently deprives the applicant, and any other person of her age, of the status of a socially insured person by making her a pensioner of the state, even if she is still looking for a new job. Regardless of her personal skills and situation, she is as it were stigmatized as belonging to the group of persons who are no longer part of the active population....

The Court's consideration of this case so soon after *McKinney* demonstrated that the specific facts of each situation need to be examined, that its earlier decision did not provide judicial sanction for mandatory retirement, and that governmental employers were required to justify such policies. The Court expressed its concern with the "insidious stereotype" of ageism whether it results in intentional ("direct") or unintentional ("adverse impact") discrimination and articulated a special concern for economically vulnerable older citizens: "The most unfortunate aspect of s. 31 is that it has the effect of denying unemployment benefits precisely to those who need them most."[48] To what extent the Court's words hold more rhetorical or substantive power is discussed below. First, we need to consider the nature of the connection between constitutional and human rights principles as they arise in jurisprudence and apply to forced retirement.

The Meiorin case, while not a mandatory retirement case, considered the basis on which discrimination cases are to be decided, emphasizing human rights language, and strengthened the BFOR standards (*British Columbia (Public Service Employee Relations Commission) v. BCGSEU* [1999]).[49] After three years as a firefighter, Tawney Meiorin failed to meet an aerobic fitness test and was consequently dismissed. Justice McLachlin, writing for the Court, overturned the Court of Appeal and upheld the arbitrator's determination to reinstate Firefighter Meiorin. In doing so, Justice McLachlin focused on the discriminatory effect of the impugned law, thus unifying Charter and human rights perspectives. She writes, "I see little reason for adopting a different approach when the claim is brought under human rights legislation which, while it may have a different legal orientation, is aimed at the same general wrong as s. 15(1) of the *Charter*."[50]

Justice McLachlin argued for a "unified approach that (1) avoids the problematic distinction between direct and adverse effect discrimination, (2) requires employers to accommodate as much as reasonably possible the characteristics of individual employees when setting the workplace standard, and (3) takes a strict approach to exemptions from the duty not to discriminate ..."[51] The emphasis on a stricter standard is a shift away from the *Large* decision, which adopted a lower standard of defence of age discrimination.[52] Justice McLachlin reinforced the importance of a strict approach by noting employers' obligations to take into account individual and group differences. "By enacting human rights statutes and providing that they are applicable to the workplace, the legislatures have determined that the standards governing the performance of work should be designed to reflect all members of society, insofar as this is reasonably possible."[53] The Court's focus seems to be shifting toward requiring employers to accommodate employees based on their right to have their human dignity acknowledged practically in employment situations. The jurisprudence in Meiorin is based in part on a case decided earlier the same year, *Law v. Canada*.

In *Law v. Canada (Minister of Employment and Immigration)* [1999],[54] Justice Iacobucci, writing on behalf the Court, focused on the goal of assuring human dignity as a way of defining the more abstract concepts of equality and discrimination:[55]

> [T]he equality guarantee in s. 15(1) is concerned with the realization of personal autonomy and self-determination. Human dignity means that an individual or group feels self-respect and self-worth. It is concerned with physical and psychological integrity and empowerment.... It is enhanced by laws which are sensitive to the needs, capacities, and merits of different individuals, taking into account the context underlying their differences.... Human dignity within the meaning of the equality guarantee does not relate to the status or position of an

> individual in society *per se*, but rather concerns the manner in which a person legitimately feels when confronted with a particular law.

Law potentially provides a new platform to revisit the issue of mandatory retirement. Justice Iacobucci writes, "The possibility of new forms of discrimination denying essential human worth cannot be foreclosed."[56] In discussion of the purpose of section 15(1), the Court emphasizes the importance of "essential human dignity and freedom" and rejects differential treatment based on stereotypical characteristics that has the effect of "perpetuating or promoting the view that the individual is less capable, or less worthy of recognition or value as a human being."[57] The Court also notes that under the Charter's equality guarantee, dignity relates not to status per se, but "concerns the manner in which a person legitimately feels when confronted with a particular law."[58] Denial of employment based on an arbitrary age (and not supported by behavioural evidence) may yet come to be seen by the Supreme Court of Canada as a form of discrimination that perpetuates and promotes the stereotype that older citizens are "no longer part of the active population."[59] In *Law*, the Court's review of jurisprudence relied on a number of cases but made only minor use of the *McKinney* decision. In spirit, if not by direct attribution, the unanimous Court in *Law* echoes the voices of the dissents in *McKinney* and of human rights legislation. The fundamental value of human dignity, which Justices Wilson and L'Heureux-Dubé emphasized in their *McKinney* dissent and which is the elemental goal of human rights codes, is given renewed consideration in *Law*. Should the Supreme Court choose to use it, in *Law* and Meiorin, the Court has given itself a platform for a rejuvenated analysis of mandatory retirement, returning to underlying principles articulated in *Etobicoke*. Largely economic concerns ought not to trump fundamental human rights.

Not only have the legal grounds shifted and re-shifted, the demographic and socio-economic circumstances and the interpretation of legislative actions have evolved and cast doubt on the permanence of the older opinions. Should the Supreme Court again rule on mandatory retirement, it might well place less weight on the socio-economic arguments of the majority in *McKinney*, and stress the human rights of older workers emphasized by the dissenting justices and more recent decisions. As a caution, not all the reasoning in *Law* will make it easier for claims of age discrimination to succeed. *Law* makes it clear that claimants bear the onus to demonstrate not only discrimination under the Charter (section 15) but also the adverse effect of such discrimination on the claimant's life.[60]

That the current judicial situation concerning mandatory retirement is ripe for change is illustrated by the fact that some lower courts and arbitration bodies are urging the Supreme Court of Canada to reconsider *McKinney*. In 2000, an arbitration board in the case of *Greater Vancouver Regional District* rejected the argument that the law was settled by *McKinney* and ruled that the Supreme Court of Canada contemplated a case-by-case determination. The arbitration award was upheld by the Court of Appeal for British Columbia in *GVRD Employees' Union v. GVRD* (2001).[61] The majority of the appellate court upheld the arbitrator's finding that the employer was required to justify its mandatory retirement policy under section 1 of the Charter, but also Justice Prowse writes: "… I would urge the Supreme Court of Canada to reconsider this issue [à la *McKinney*]…. The extent to which mandatory retirement policies impact on other equality rights, and on the mobility of the workforce, have become prominent social issues."[62]

A Social Analysis of the Judicial Context: A Renewed Emphasis on Human Rights?

We argue that in the last quarter century, the Supreme Court of Canada has done something like a full 360° turn in the development of the jurisprudence on mandatory retirement. The shifts in the Court's position are summarized by the cases discussed here. In *Etobicoke* (1982), the Supreme Court rejected the legality of compulsory retirement at age sixty that was provided for in a collective agreement. The Court argued that individual capacity had to be considered seriously. The majority opinion in *McKinney* (1990) marked a substantial change from this earlier opinion. In *McKinney*, the majority developed a socio-economic argument that mandatory retirement at age sixty-five was justified as part of the fabric of work arrangements in Canada and that age discrimination did not deserve the same level of legal scrutiny as other forms of discrimination. The dissents in *McKinney* emphasized that section 15 of the Charter was intended to promote human dignity. Not for the first time in judicial history, the language of the dissents may have a longer life than the perspective of the majority. Thus, in *Law* (1999) the court reiterated the importance of individual ability by noting that "a focus is quite properly placed upon the goal of assuring human dignity by the remedying of discriminatory treatment."[63] In this way, the Supreme Court spirals back toward the position in *Etobicoke*.[64]

The Charter constrains the power of the state, and human rights legislation regulates the actions of private parties. In *McKinney*, the Court clearly separated its analysis of the constitutionality of mandatory retirement and the constitutionality of the Ontario Human Rights Code. The Supreme Court has said, for example in *Dickason*, that while deference is due legislation, private employers' policies and practices are not entitled to the same respect. The jurisprudence on mandatory retirement subsequent to *McKinney* has brought these differentiated strands of legal analysis closer together. Throughout most of the 1990s, the Supreme Court of Canada argued that age discrimination did not deserve the same level of protection as other designated groups. By 1999, however, the Court's decisions showed greater concern for ageism, and it moved toward an analysis of discrimination that integrated Charter and human rights approaches in at least two ways. Although the Supreme Court has not overturned its earlier rulings on mandatory retirement, it has strengthened the BFOR test; and the courts moved away from a socio-economic analysis of forced retirement toward a human rights analysis.

The extent to which the Supreme Court of Canada will give greater weight to human rights when considering future involuntary retirement cases remains an open question. Despite the Court's expressions of concern about ageism (for example, in *Tétreault-Gadoury*), questions remain about its determination to redress human rights abuses in the private sector. For example, Anderson argues that the Court's acceptance of collective agreements as probative of the reasonableness of a challenged practice gives deference to private policies at the expense of human rights. The human rights codes "do not 'fill the *Charter* gap' because they fail to implement a more social democratic vision of human rights."[65] Those aged sixty-five and older constitute an economically disenfranchised minority with fewer statutory protections than younger employees. This systemic issue, inadequately addressed by legal consideration, treats age discrimination as an aberrant individual matter. A whole class suffers—often overlapping with other categories, such as gender and ability.[66]

More broadly, Beatty argues that judges are open to the criticism that they favour employer and investor interests over those of employees, that "many of the most important interests and activities of workers—including the right to strike, picket, select a union, be free from discrimination, and not to be terminated (retired) against their will—effectively enjoy none of the protection that many thought the Charter provided."[67] To the extent that employers are not required to consider alternatives to their indirect discriminatory policies, especially where it would be relatively easy to do so, human rights codes will be ineffective.[68] As Crane notes, one elemental goal of human rights legislation is to promote inclusion and equal participation of diverse groups in the workplace and the duty to accommodate is one of the most effective tools available to ensure these rights.[69]

The principles of human rights constantly challenge society to rethink the assumptions underlying the organization of work at any given time. For today's needs, the majority opinion in *McKinney* only inadequately meets this challenge; consequently, it has been a lightning rod for criticism, was effectively de-emphasized in *Law*, and seems unlikely to be the basis of a final resolution of the mandatory retirement issue.

Legislative and Policy Developments

Along with the shifts and counter-shifts of jurisprudence in the past two decades, considerable developments have occurred in the past five years in the legislative, political, and policy arenas. We briefly examine four sets of recent changes that influence mandatory retirement: (1) recommendations of human rights commissions and civil society groups, (2) positions taken by political parties and political leaders, (3) labour market trends, and (4) the experience of other Anglo-Saxon liberal market economies.

Increasingly, legal attention has been given to the assumptions and implications of human rights legislation. In 1999, the federal Minister of Justice established an independent panel to conduct a review of the Canadian Human Rights Act, chaired by retired Justice La Forest, who wrote the majority opinions in *McKinney* and related decisions. The recommendations of the panel, released in 2000, included that a review be undertaken of mandatory retirement "in the federal sector based on human rights principles and socio-economic factors, to determine whether mandatory retirement should be subject only to the BFOR or whether more specific defences should be crafted to allow for mandatory retirement in defined circumstances."[70] Additionally, the panel recommended "there be no blanket defences for mandatory retirement."[71]

Also in 1999, which was the International Year of Older Persons, the Ontario Human Rights Commission began a consultation process on age discrimination. In the words of one human rights official, the consultation "touched something" with the response from the public being far greater than anticipated. The result of the consultation was the 2001 report, *Time for Action: Advancing Human Rights for Older Ontarians.* The report was "a broad examination of all issues that may have an impact on the dignity and worth of older adults and that may affect the enjoyment of equal rights and opportunities."[72] It emphasized the inter-connection of age with other designated categories such as gender, disability, sexual orientation, immigration status, race, and ethnicity.

Specifically with respect to mandatory retirement, the report recommended:[73]

> That the *[Ontario Human Rights] Code* [1990] be amended to eliminate the blanket defence to mandatory retirement at age sixty-five and to extend protection against age discrimination to workers over sixty-five. This could be done by removing the upper limit of sixty-five in the definition of "age" in section 10(1). Employers who wish to have age-based retirement policies will be required to demonstrate that the policy is based on *bona fide* occupational requirements. Laws and programs that require consequential adjustment should also be reviewed.

> That, irrespective of whether the *Code* is amended, employers and unions reconsider the utility and necessity of requiring employees to retire at age 65 and revise their retirement policies and collective agreements to promote flexibility and choice.

Although no legislative changes have yet resulted from the two human rights reports, a political debate on mandatory retirement has also recently emerged as reviewed by Gillin (Chapter 1). The current demographic and labour market conditions are considerably different from those at the time of *McKinney* in 1990 and in the 1980s, as outlined by Kerr and Beaujot (Chapter 6) and discussed in several other chapters of this volume.

Conclusions

Greene ends his book on the Charter by noting, "[T]he measure of our commitment to human rights is the degree to which we believe in and practice interpersonal respect, the extent to which we are all able to achieve a sense of self-worth and encourage it in others, and the success we have in treating each other fairly."[74] The sequence of the Supreme Court decisions analyzed here challenges us to consider whether the practice of involuntary retirement at age sixty-five adequately reflects the Canadian community's commitment to recognizing self-worth, and whether this practice succeeds in establishing fairness in the lives of all.

In decisions reached during the early 1980s, the Supreme Court treated age as an insufficient ground for justifying mandatory retirement, whereas in *McKinney*, and other cases decided in 1990, age becomes a legitimate ground for achieving certain management objectives. In *McKinney*, the Court retracted the broad and purposive interpretation of equality found in the earlier decisions and replaced it with a position more reflective of classical liberalism and more protective of the economic interests of the status quo than community values. In its mandatory retirement decisions under the Charter, the Supreme Court displayed a disturbing tendency to articulate principled standards to establish the reasonableness of section 1 limits, then to slide past these standards without addressing them in a convincing manner. In a previous analysis of *McKinney*, we suggested that it was likely the judges would prefer the legislators to decide mandatory retirement issues.[75] Indeed, in *McKinney* the Court deferred to legislators; though on other issues, such as those dealing with sexual orientation and employment, it has not done so. Nonetheless, this chapter proposes that the Supreme Court appears to have reconsidered its *McKinney* decision and repositioned itself to emphasize that equality and dignity are paramount, rather than workplace practice or labour market conditions. This repositioning has yet to be fully applied to mandatory retirement, and there is no guarantee that it will be. The current Court seems to select very carefully where it chooses to intervene and tends to pursue a pragmatic approach.[76]

In 1999, the Court moved away from its earlier and sociologically flawed argument in *McKinney* to a sounder legal argument in *Law*, emphasizing human rights and the "needs, capacities and merits" of employees.[77] The *Law* decision also might be read as a philosophical shift in another sense. The increase in the development and influence of human rights legislation reflects the growing awareness that not only can government institutions pose a threat to personal autonomy, but also powerful private interests, such as large employers, can systemically and unfairly limit employee rights and freedoms. The arbitrariness of forced retirement at a specific age is no less repugnant than the arbitrariness of denying employment opportunities on such bases as gender, race, colour, religion, or sexual orientation. In the same year, in the Meiorin case, the Court adopted a more stringent BFOR standard that places a greater onus on employers to justify why individuals should not be accommodated.

These concerns echo the tone and substance of the dissents written by Justices Wilson and L'Heureux-Dubé in *McKinney*. The equity promise of the Charter is to protect against prejudice, stereotype, and discrimination—to promote human dignity.[78] Justice Wilson writes, "[T]he rights of the appellants which have been infringed pertain to their dignity and sense of self-worth and self-esteem as valued members of the community, values which are at the very centre of the Charter."[79] As discussed earlier, these concerns are also central to the arguments of human rights advocates. By its nature, the law builds upon its past constructions and extends them into the future by making an authoritative claim upon us.[80] The Canadian community ought to promote the worth and human dignity of workers sixty-five and older who have or seek paid employment by eliminating mandatory retirement. This would be most consistent with values inherent in the Charter and contemporary jurisprudence.

The literature in support of mandatory retirement presents two basic types of arguments why courts should not seek to alter existing arrangements. First, as expounded by Gunderson and Hyatt (Chapter 8), from a market perspective, mandatory retirement is freely entered into and has utility to employers and workers. The difficulty with this argument is that it fails to give appropriate weight to human rights. Historically, workplace policies—even those that might have been "freely bargained"—incorporated systemic discrimination. As a result, pay and employment equity, as well as human rights legislation—all subject to judicial review—were established to redress such discrimination. Secondly, Hogg proposes, "[J]udicial intervention in the complex field of employee retirement is bound to be clumsy, and is likely to produce counterproductive results."[81] However, whatever the originating source of intervention, changes to the social structure are inevitably complex, difficult to fully implement, and have unintended consequences. The complexity argument has been used in the past to rationalize race, gender, and other inequalities. In any case, with regard to mandatory retirement, the experience of Quebec, Manitoba, the United States, Australia, and New Zealand provide no support that eliminating mandatory retirement policies is unduly disruptive.

The law has powerful constructive opportunities and obligations to further human rights, but the responsibility is not the law's alone. Under the Constitution and the Charter, the courts have specific duties to ensure fairness and equity. In earlier eras, Canadians accepted discrimination in hiring and wages based on race, sex, and marital status. It may be that some are prepared to tolerate ageist laws in a way that is now unthinkable with regard to racism or sexism, but we think the drive for equality will inevitably encompass age as well. Over the past several decades, the worst forms of racism, sexism, homophobia, and discrimination against persons with disabilities have been challenged and significantly lessened—though hardly eliminated. This has been the result of judicial, legislative, and political action. It is past time to address the adverse effects of ageism.

At the beginning of the chapter, we proposed that fundamental law is about envisioning a community. To realize the vision of the Charter and human rights codes requires social involvement as well as legal conceptualization.[82] Currently, the mandatory retirement regime in most parts of Canada is a form of legalized discrimination. What is missing from the existing jurisprudence is the *explicit* acknowledgment that to speak of a "normal" retirement age, as the majority Justices did in *McKinney*, is to treat forced retirement *as if* it should be taken for granted. To conceive mandatory retirement as part of the fabric of Canadian society is to affirm ageism; it legitimates systemic discrimination. Workplace management schemes—whether derived from collective agreements, pension arrangements, or even the legislated human rights codes—ought not to trump the fundamental human rights enshrined in the Charter. Overcoming ageist laws is an extension of the historical efforts to forge a more just society. The idea of equality and dignity across human ages will become a political reality when Canadians invest their energies in building this kind of community.

Note: Some data for this chapter were obtained from interviews with officials of human rights commissions, unions, and law firms. We thank Patrick Dickie for his insightful comments on an earlier draft of this chapter, and David Forgione for his research assistance. Thomas Klassen acknowledges the financial support of a grant from the York University Centre for Research on Work and Society as part of its Social Sciences and Humanities Research Council (SSHRC) Initiative on the New Economy Research Alliance project. C.T. Gillin thanks Ryerson University and his colleagues for their support of his scholarship.

Cases Cited

Andrews v. Law Society of British Columbia, [1989] 1 S.C.R. 143.

British Columbia (Public Service Employee Relations Commission) v. BCGSEU, [1999] 3 S.C.R. 3.

Dickason v. University of Alberta, [1992] 2 S.C.R. 1103.
Douglas/Kwantlen Faculty Association v. Douglas College, [1990] 77 D.L.R. (4th) 94.
GVRD Employees' Union v. GVRD, [2001] BCCA 435.
Harrison v. University of British Columbia, [1990] 77 D.L.R. (4th) 55.
Irwin Toy Ltd. v. Québec (Attorney General), [1989] 1 S.C.R. 927.
Large v. Stratford (City), [1995] 3 S.C.R. 733.
Law v. Canada (Minister of Employment and Immigration), [1999] 1 S.C.R. 4.
McKinney v. University of Guelph, [1987] 46 D.L.R. (4th) 193. (Ontario Court of Appeal)
McKinney v. University of Guelph, [1990] 3 S.C.R. 229. (Supreme Court of Canada)
Ontario Human Rights Commission v. Borough of Etobicoke, [1982] 132 D.L.R. (3d) 14.
R. v. Oakes, [1986] 1 S.C.R. 103.
Re Winnipeg School Division No. 1 v. Craton, [1985] 21 D.L.R. (4th) 1.
Ref. Re Public Service Employee Relations Act, [1987] 1 S.C.R. 313.
Rodriguez v. British Columbia (Attorney General), [1993] 3 S.C.R. 519.
Stoffman v. Vancouver General Hospital, [1990] 76 D.L.R. (4th) 700.
Tétreault-Gadoury v. Canada (Employment and Immigration Commission), [1991] 2 S.C.R. 22.
Vriend v. Alberta, 1 S.C.R. [1998] 493.

Appendix: Canadian legislative provisions on mandatory retirement age by federal, provincial, or territorial jurisdiction

Canada	Mandatory retirement is not a discriminatory practice when a person has reached the normal retirement age for employees performing the same type of work. Consequently, in that case, the Act allows for mandatory retirement.
Alberta	Mandatory retirement constitutes a discriminatory measure for employers under the jurisdiction of this province.

British Columbia	Older employees are protected until the age of sixty-five against discrimination based on age. Consequently, employees aged sixty-five or over cannot file a complaint if they are obliged to retire for this reason.
Manitoba	Mandatory retirement constitutes a discriminatory measure for employers under the jurisdiction of this province.
New Brunswick	Termination of employment provided for in a retirement or pension plan does not constitute a discriminatory measure. In the absence of such a plan, however, employees who are obliged to retire may file a complaint for discrimination based on age under the legislation on human rights.
Newfoundland and Labrador	Termination of employment provided for in a retirement or pension plan does not constitute a discriminatory measure. In the absence of such a plan, however, employees who are obliged to retire may file a complaint for discrimination based on age. They may use this recourse until the age of sixty-five.
Northwest Territories	Mandatory retirement constitutes a discriminatory measure for employers under the jurisdiction of this territory.
Nova Scotia	Mandatory retirement at age sixty-five does not constitute a discriminatory measure if it is standard in the workplace in question. However, the Human Rights Commission of this province investigates when an employee aged sixty-five or over is not treated in the same manner as others of the same age where retirement is concerned.
Nunavut	Mandatory retirement constitutes a discriminatory measure for employers under the jurisdiction of this territory.

Ontario	Older employees are protected against age-based discrimination up to the age of sixty-five. Consequently, employees aged sixty-five or over cannot file a complaint if they are obliged to retire for this reason.
Prince Edward Island	Mandatory retirement constitutes a discriminatory measure for employers under the jurisdiction of this province.
Quebec	Mandatory retirement constitutes a form of discrimination according to the Charter of Human Rights and Freedoms and, more explicitly, is forbidden by the Act Respecting Labour Standards.
Saskatchewan	Older employees are protected against age-based discrimination up to the age of sixty-five. Consequently, employees aged sixty-five or over cannot file a complaint if they are obliged to retire for this reason.
Yukon	Mandatory retirement constitutes a discriminatory measure for employers under the jurisdiction of this territory.

Source: HRDC Web site: *http://labour.hrdc-drhc.gc.ca/worklife/aw-retirement-legislative-02-en.cfm* (accessed October 15, 2003)

4

Forced Retirement: Organized Labour's Predicament

Thomas R. Klassen with David J. Forgione

Retirement, we understand, is great if you are busy, rich and healthy. But then, under those conditions, work is great too.

Bill Vaughan[1]

Retirement—whether forced or not—is a complex social institution in contemporary Canada. Retirement represents a significant event for individuals, vis-à-vis employment and labour force attachment, but more fundamentally it is a primary transition point in the life cycle. In modern industrial societies, retirement and grandparent hood have traditionally been the markers of old age.[2] Retirement, particularly mandated retirement, also illustrates the complex relationships that have come to exist between state, labour, and capital in the twentieth century.

This chapter examines the workplace dynamics that have shaped the institutional order surrounding compulsory retirement, with a focus on the role of organized labour and the position(s) it has taken on compulsory retirement outside of Quebec.[3] The arbitrary withdrawal of the ability, or right, to work would seem to be anathema to labour unions in a capitalist society. Yet, as will become apparent, this is not necessarily the case. As there has been no previous research on the role of Canadian unions in the mandatory retirement debate, this chapter represents the first examination of how unions are reacting to current debates.

Compulsory retirement is a creation of contemporary capitalism, in which most workers in Canada (85 percent) have retirement age dictated or must negotiate this individually with an employer or through collective agreements reached by labour unions. Indeed, age sixty-five as the age of forced retirement is even more recent. Canada's Old Age Pension Act, in force from 1927 to 1952, set the pensionable age at seventy, rather than sixty-five as was the case in most other developed nations at that time.[4] Even in the 1950s, under the Old Age Security Act, and in 1965, when the Canada and Quebec Pension Plans were introduced, age seventy was set as the age when recipients became eligible for full (non-reduced or means-tested) benefits.[5] Only in the late 1960s did age sixty-five became the accepted marker for exit from the labour market and entry into old age—primarily because federal income support programs began to adopt this age as the trigger for full entitlement to the Old Age Security, the Guaranteed Income Supplement, and the Canada Pension Plan.[6] At present, a range of other legislation and public policies, including taxation, health care and human rights, support the institutionalization of age sixty-five. Employers and unions have incorporated this age into collective agreements and other workplace arrangements, especially with regard to private pensions.

The institution of retirement at a predetermined age, along with a guarantee by the state (and to some degree employers) of a minimum level of income, is one of the monumental accomplishments of the twentieth-century labour movement. The income security provisions gained by retired workers in the past fifty years have, to date, proven remarkably resilient. Although other components of the Canadian welfare state such as health care and social assistance have been subject to considerable pressure over the past three decades, income security programs for those over age sixty-five have been among the least affected by the retrenchment of the Canadian state.

State, Capital, and Labour

The central actors in shaping both the existing mandatory retirement arrangements and future developments are: (1) the state, (2) employers, and (3) workers. In examining the state, two groups will play key roles in changing existing policies and laws. Politicians may seek to alter existing arrangements in response to shifts in demographic conditions and associated voting patterns, in combination with pressures from capital, labour, or other groups. The second state actor is the judicial system in that courts, especially in interpreting the Canadian Charter of Rights and Freedoms, might rule that the existing arrangements are contrary to the fundamental rights guaranteed to citizens.[7] Of course, the dynamic of state action—or inaction—in regulating the labour market is a complex one, particularly so in a decentralized federation such as Canada.[8] The state is also an employer, and moreover, there is considerable debate about its scope for action that is independent of capital. The existence of private property and the dependence of the state on capital have always imposed significant boundaries on democratic politics and policy-making.[9]

Within the state, at least two major cleavages are visible with respect to forced retirement. First, the Charter of Rights and Freedoms through section 15 grants a set of fundamental equality rights, including the right not to be discriminated against due to age. However, the Charter's approach of granting equality, especially with regard to the employment relationships, runs counter to the common law tradition of English Canada in which employment contracts are private and individual arrangements. As such, as noted by Gillin and Klassen (Chapter 3), the Supreme Court has typically remained conservative in rulings that might favour workers or extend the rights of organized labour.[10] In any case, labour has also often sought to exclude judicial involvement, preferring instead to rely on collective agreements to regulate capital.

Second, the introduction of the Charter has created an even more multifaceted relationship between Parliament and the courts by expanding the scope of judicial review. In certain cases, such as abortion, sexual orientation, and same-sex rights, the Supreme Court has been activist in reviewing, if not overturning, laws passed by legislatures.[11] According to some observers, activist judges display "a readiness to veto the policies of the other branches of government on constitutional grounds, usually by striking down statutes."[12] Cumulatively, the outcome of these decisions has somewhat refashioned the relationship of the courts to the legislative and executive branches of government. This has not, however, been the case with regard to mandatory retirement, especially in the 1990 *McKinney* landmark decision in which the Supreme Court ruled that although mandatory retirement, as embodied in provincial statutes, is age discrimination, it is nonetheless constitutional for some workers.[13]

Employers are the second major group that might be expected to seek to pressure mandatory retirement regimes. As with the state, the dynamic of employer behaviour is complex and is a function of the specific environment faced by an employer: degree of competition, nature of the industry, labour market conditions, extent of unionization, wage levels, and so forth. Within capital, there are divisions on the matter of when and how workers should retire, or perhaps more correctly, should be retired. On the one hand, employers should be expected to support mandatory retirement on the basis that it provides certainty on staffing and ensures that higher-paid older workers are eliminated from the labour force. Indeed, this is the position taken by many employers and employer associations supporting a set retirement age of sixty-five, and reviewed in general by Gunderson and Hyatt (Chapter 8) and Kesselman (Chapter 9), and with regard to universities by other contributors to this book. However, employers could also be logically expected to support the elimination of mandatory retirement given that doing so would provide them with greater flexibility in the allocation of human resources.

Workers, the party most immediately affected by mandatory retirement, would seem to be the actor that would be most active in debates about the regulation of departure from paid employment, and the group that would expend the most effort to solidify its position.[14] Yet, here too, the situation is complicated and by no means stable. For instance, this institution may not directly affect some workers, especially self-employed or contract workers, as well as those in jurisdictions or organizations that have eliminated mandatory retirement. In fact, most workers prefer early retirement, and given that the average retirement age was 61.2 years of age in Canada in 2002, it seems that many workers do not face compulsory retirement at age sixty-five.[15] A number of factors—education, type of occupation, income, health status, family structure, etc.—affect individual decisions and attitudes about mandatory retirement. Sharp divisions and contradictions exist within the labour movement between those unions that support mandatory retirement policies and those that do not.

In North America, immediately after the Second World War, labour unions generally opposed mandatory retirement.[16] However, a general trend emerged in that as pension benefits (and retirement incomes) increased, so did organized labour's acceptance of automatic retirement schemes.[17] This was the pattern in Canada as well; as the welfare state expanded, and pensions became institutionalized, unions increasingly supported mandatory retirement.

Union Positions

Many Canadian unions, especially the national and provincial offices, support mandatory retirement on three grounds. First, compulsory retirement at a fixed age creates jobs and promotions for younger workers, and thereby reduces the unemployment rate. This view is shared by some economists who see mandatory retirement "as a form of life cycle work sharing, whereby older workers vacate jobs that create promotion opportunities for middle-aged workers. Their promotion, in turn, creates new job opportunities for younger workers which may also lead to an infusion of new ideas and talents into the organization."[18] Gunderson and Hyatt (Chapter 8) and Kesselman (Chapter 9) review this contention in more detail.

Second, unions support mandatory retirement because without it, governments and employers are thought to have greater scope to reduce pension benefits, which would force workers to remain employed for longer. For decades, the labour movement has sought to reduce the working life of workers and enhance pension benefits. As such, any possibility that these gains may be eroded logically will be resisted.[19] Third, unions support compulsory retirement because it reduces the scope for employer monitoring and evaluation of older workers by setting a binding date of retirement. In other words, fixed retirement protects less productive workers because employers, particularly in a unionized workplace, are willing to tolerate marginal performance to the date of mandated retirement.[20]

The Canadian Labour Congress (CLC), the major coordinating body for unions in Canada, representing 2.5 million workers, has, since at least 1980, formally endorsed mandatory retirement. Furthermore, the constitution of the CLC prevents members from running for office if they are sixty-five or older, and in 2003, after a bitter strike, the Congress imposed compulsory retirement at age sixty-five for its own staff. The president of the CLC, in public pronouncements, has focussed on the pension argument by stating that employers and the government "haven't funded our pension plans properly … and the easiest way for them to get out of their obligations is to work us till we die. Then they don't have to pay pensions."[21]

Nevertheless, the Congress's own analysis concludes that banning mandatory retirement, "on its own, has no measurable effect on the age of retirement."[22] Furthermore, CLC staff has concluded that the argument that compulsory retirement will create jobs for young workers is both "bad economics and dangerous ethics."[23] It is seen as bad economics because the number of jobs in the economy is not fixed, and dangerous ethics because it results in creating "job opportunities for those we favour (young workers) by getting rid of those we don't (older workers)."[24] As such, the CLC's support for mandatory retirement seems to rest primarily on the fact that unions accept "it as a *quid pro quo* for other things—especially pension benefits."[25]

The union most strongly opposed to the changes in mandatory retirement provision is the Canadian Auto Workers (CAW), the nation's largest private sector union. The CAW's formal position is premised on forced retirement as a mechanism for organizational renewal. Adopted in 1997, it reads in part:

> The continuing strength of our union is related to our ability to bring in new activists, shop stewards, and leadership ….
>
> Many of our major collective agreements, covering large numbers of our members, now require compulsory retirement at age sixty-five.

> As we head toward the new millennium, this convention strongly supports the move, from a political decision, to a mandatory age sixty-five retirement provision in the constitution of our union where the law allows. In those jurisdictions where the law does not allow mandatory retirement at age sixty-five, this Convention statement shall serve notice to all national officers, as well as local union officers and staff, that the expectation and desire of the union is that all leadership voluntarily abides by this convention decision.[26]

Thus, the organizational renewal and promotional opportunities of concern to unions are not necessarily only those of employers, but also include those of the labour movement.

The Canadian Union of Public Employees (CUPE) has similar concerns to those of the CLC about the removal of the mandatory retirement age, especially related to pensions. As Sid Ryan, president of CUPE Ontario noted, "Eliminating mandatory retirement is part of the harmonization of social and pension programs that has occurred since the free trade debate in 1988. It is an opportunity to undermine existing pension schemes. This is the danger it brings about."[27]

That a central plank in the labour movement's position on mandatory retirement is related to private pension schemes requires elaboration. There are two major types of pensions: defined benefit and defined contribution. In defined benefit arrangements, the benefits paid relate to final salary (or a proxy thereof) and years of service. In such schemes, individual workers gain disproportionately from contributions made close to retirement. Younger workers can be treated in this unfair manner because the plans are structured on the assumption that workers will be long-term members of the scheme. Fixed and predetermined retirement ages are critical in such plans because typically it is impossible for workers to accrue further benefits once reaching the normal retirement age. In other words, continuing to contribute to the plan typically causes the loss of the immediate payment of benefits as well as not increasing the level of the benefits to be paid in the future.[28]

In recent years, many employers have sought to shift pension schemes toward the defined contribution model.[29] Under this model, pension benefits of workers are related to the level of assets they have accumulated in the fund, as well as the number of years over which benefits will have to be paid. In other words, the benefits paid to workers are calculated at the date of retirement and are based on accumulated contributions and investment income. As such, the age of retirement is less important in the operation of the pension plan because essentially the plan sets no predetermined level of benefits. Workers who contribute to the plan longer (even past the normal retirement age) draw a higher payment than those who retire earlier, not only because they contribute longer but also because the pension will have to be paid for fewer years.

Defined benefit pension plans are much preferred by unions, as they provide workers with better security and higher benefits. The employer bears greater risk in ensuring that the plan is financially sound to pay out the predetermined benefits to retired workers. Additionally, defined benefit plans financially reward workers who remain with the same employer, which is advantageous to unions in retaining members. In defined contribution plans, risk is shifted to workers, who remain uncertain about the performance of the investments that constitute the pension plan and the level of benefits to be paid. It is the less desirable defined contribution plans that are more compatible with removing a fixed retirement date. Not surprisingly, pension issues are at the forefront for unions when mandatory retirement is examined.

Notwithstanding the positions taken by the national and provincial offices of unions, Canada's decentralized labour movement allows individual workplace union locals to largely ignore the central offices. At the local level, CAW, CUPE and other unions have a more nuanced perspective because local offices sometimes include members who want to continue to work past the mandated retirement age. Union locals are known to launch grievances in support of workers who do not wish to be retired at an arbitrary age, even if doing so is inconsistent with national office policy.

The pro-mandatory retirement stance taken by the CLC does not bind its members, and as such, a second position exists within the labour movement—namely, for unions to have *no* formal policy on mandatory retirement. For example, the United Food and Commercial Workers Union (UFCW) has no policy on mandatory retirement, either as an employer or for its locals, and does not oppose legislative and policy shifts to eliminate mandatory retirement. As an employer it encourages, but does not force, staff to retire by age sixty-five. Typically, unions with no formal position on mandatory retirement are unwilling (or unable) to obtain support from their membership to challenge mandatory retirement regimes, but at the same time subscribe to non-discrimination policies.

Despite the lack of a position, or with mutually contradictory or incompatible positions, unions in this group also emphasize the importance of pensions vis-à-vis retirement. For UFCW, issues related to mandatory retirement are directly related to adequate pension benefits and the shift from defined benefits to defined contribution schemes. The union, which typically represents workers with low pay, is concerned that eliminating compulsory retirement, combined with decreases in pension benefits, will result in some workers having no choice but to work past what has been the traditional retirement age.

In an effort to make certain that workers are entitled to the protection granted by such legislation, many unions, especially in the education and health care sectors, in British Columbia (and other provinces) have ensured, often unwittingly, that mandatory retirement at age sixty-five is effectively incorporated into collective agreements.[30] They may oppose forced retirement in principle, but have locked themselves into a collective bargaining regime that validates mandated retirement. For other unions, reaching a formal position on mandatory retirement has been a low priority. The formally democratic nature of unions means that the views of the majority (who wish to retire early with high pension benefits) take precedence over the fewer members who wish to work past the normal retirement age.

The largest union in Canada whose members are not forced to retire, the Public Service Alliance of Canada, which represents federal civil servants, has yet to arrive at a formal position on compulsory retirement. This is, at least in part, because the largely unprompted elimination by the federal government of mandatory retirement for its workers in 1986 has had no discernable impact. In fact, the average age of retirement of federal civil service workers has declined somewhat since.

In addition to the unions opposed to changes in mandatory retirement laws and policies, and those that lack a clear stance, the third position within the labour movement with respect to forced retirement is found in a smaller group of unions, especially those representing professional and highly paid workers. These unions oppose compulsory retirement, usually on the basis that it is discriminatory and violates fundamental human rights. In this regard, the Canadian Association of University Teachers (CAUT), although not strictly a union but rather a federation of faculty associations and unions, is instructive. Its position on mandatory retirement is that:

> Mandatory retirement is discrimination on the basis of age, and may give rise to discrimination on the basis of sex or other grounds. Academic staff have a right to continue their employment beyond the standard retirement age under the same terms and conditions.[31]

The federation has not always held this position. In fact, until 1979, it did not identify mandatory retirement at an arbitrary age as a discriminatory practice.[32]

One provincial counterpart to the national association, the Ontario Confederation of University Faculty Associations, also has a policy opposing forced retirement, and was one of the organizations that supported the *McKinney* Supreme Court challenge in 1990. Like CAUT, it conceptualizes compulsory retirement as primarily an infringement on the human rights of workers. However, as with the unions opposed to changes to current arrangements, the support for change of the national and provincial federations does not mean that all individual faculty unions have a similar position.

Analysis

In reviewing the stance of organized labour outside of Quebec with regard to forced retirement at age sixty-five, it appears unlikely that unions, especially the large and powerful ones, will support the elimination of mandatory retirement. Indeed, the opposite is the case. Yet at the same time, a group of unions without a clear stance and a small set of organizations within the labour movement oppose forced retirement. The analysis below shows why the bulk of organized labour continues to support mandatory retirement when it is an ageist and discriminatory practice.

Many unions do not frame mandatory retirement as primarily a human rights or age discrimination issue. Union leaders and many members continue to view mandatory retirement as a collective right of workers attained over many decades. A central feature of Canadian unions is the collective agreement, which binds employers to treat all their workers in a clearly circumscribed manner. As such, a tension exists between the individual rights and protections granted to workers (and to citizens under human rights legislation) and the group rights inherent in collective agreements.[33] CUPE's Ryan summarized the tension as follows:

> Certainly, discrimination on the basis of age is a legitimate human rights claim. But as trade unionists we must fall on the side of collective rights because it provides the maximum benefit to the maximum number of people. Siding with individual rights over the collective rights means making life worse for a larger number of workers than it would make life better for. In this instance, the collective right clearly outweighs the individual right.[34]

The lack of attention to individual human rights is reinforced by the Ontario Federation of Labour. It has proposed that should mandatory retirement be eliminated, the government must "ensure that current collective bargaining rights are protected; specifically *the right for workers through their union and employers to negotiate mandatory retirement*."[35]

In the history of Canadian organized labour, minority groups have not always been well served. Unions, as democratic organizations, represent the views and interests of the majority. As such, unions historically were not at the forefront of advancing the position of women in the labour force.[36] Nor have unions typically been in the vanguard of seeking to improve the labour market position of other minority groups: visible minorities, persons with disabilities, Aboriginals, and others (such as homosexuals). Workers wishing to work past the normal retirement age are a minority and hence have received similar treatment. Moreover, since almost by definition many of those over sixty-five are no longer members of unions—most voluntarily, but some not—unions are unlikely to take heed of their wishes. Union leaders will often favour the interests of the younger or middle cohort members (who are more likely to favour mandatory retirement) as this group will be union members longer than older workers.[37]

The groups most affected by mandatory retirement are those with relatively little influence in the union movement: women and immigrants, often visible minorities. Canadian unions have tended to reproduce male privilege, while assuming a largely white male membership employed full-time.[38] As Pupo and Duffy (Chapter 7) and McKenzie Leiper (Chapter 11) demonstrate, the labour force experience for many women is dramatically unlike that of men, with more women employed part-time, at lower wages, and with employment interruptions.

As noted previously, an arbitrary retirement age assists unions in dealing with the democracy dilemma—namely, that unions represent members in bargaining with capital and in relations with government, while at the same time also acting as a vehicle for democratic self-determination. The first role necessitates clear policy positions and an oligarchic structure, while the second requires participatory processes and attention to differences between members.[39] In dealing "effectively with issues that threaten members' interests, unions may be forced to narrow the range of discussable alternatives to those that, if implemented, preserve the union's strength and vigour as an institution in the long term rather than those that individual groups of workers may desire in the short term."[40]

Compulsory retirement has served the internal democratic needs of unions well. Organizationally, mandatory retirement has allowed for a degree of renewal and democracy, since long-serving elected and appointed officials must step down at age sixty-five. As first pointed out by Robert Michels in his seminal study of democracy, the tendency, indeed necessity, for democratic organizations is to become increasingly become bureaucratized with the attendant creation of an oligarchy.[41] As illustrated by the CAW's formal policy, adopting mandatory retirement was a means to ensure that its own leadership is renewed and oligarchic tendencies minimized. More broadly, both for unions and employers, mandated retirement serves as a counterbalance to the seniority principle that is inherent and fundamental to most unions.[42] As the length of service increases, workers acquire both greater protection from layoff and other benefits as well, but fixed retirement places a limit on the acquisition of seniority rights.

Unions are essentially conservative organizations that have created, in part, the institution of mandatory retirement from which many members currently benefit. The right for workers to withdraw from the labour force with a degree of income security provided by private and/or public pension was one of the battles fought increasingly by unions after the 1950s. As such, perhaps not surprisingly, the elimination of the right to a fixed and predetermined retirement age is something that unions (and the New Democratic Party) are hesitant to abandon. Indeed, many unions have exploited the 1990 *McKinney* judicial decision by claiming that it made mandatory retirement legal. However, that ruling was a narrow and highly specific one that applied only to a very small number of occupations.

Many unions view the elimination of forced retirement as yet another tactic by employers and the state in the neo-liberal era to increase the flexibility of the workforce and ultimately to reduce pension and other income security programs. Fixed retirement at age sixty-five has been an effective tool utilized by unions to press for adequate pensions from employers, and for state programs for older citizens. Without this lever, unions may be obliged to discard an important plank of their collective bargaining strategies.

The linkage of income support and mandatory retirement, critical for unions, warrants closer examination, especially in a cross-national context. In the United States, mandatory retirement before age seventy was prohibited in 1978 and then banned altogether in 1986. Although no workplace or labour market problems have been reported in association with the elimination of mandatory retirement, the United States is gradually increasing entitlement to full social security benefits to age 67 from 65. Furthermore, there are more than twice as many workers in the United States over the age of 65 as there are in Canada. In 1999 in the United States, 8.9 per cent of individuals 65 and older were in the labour market compared to only 3.4 percent of Canadians.[43] More older workers and less generous social security is what many in the labour movement predict for Canada, if mandatory retirement is eradicated. There is, however, an noteworthy difference between the two countries: the unionization rate has decreased in the United States from 20.1 percent in 1983 to 12.9 percent in 2003,[44] while the unionization rate in Canada has remained relatively constant at about 30 percent during the same period.

Australia and New Zealand largely banned compulsory retirement during the 1990s. As with the United States, no labour market or workplace dysfunctions have been reported as a result. However, the changes occurred during a time when the unionization rate in Australia plummeted from 42 percent in 1988 to only 23 percent in 2003.[45] In New Zealand the unionization rate plunged dramatically from 35 percent of the labour force in 1991 to only 17 percent by 1999, with a slight recovery to 22 percent by 2002.[46] As in the United States, the labour movements in both countries, facing severe retrenchment during the 1990s, did not play an active role in the debates about ending compulsory retirement.

It is, however, the experience of Quebec, not of other nations, that provides the most useful model for unions in other parts of Canada. Quebec eliminated compulsory retirement more than two decades ago, but its unionization rate has remained considerably higher than the national average[47] while the median age of retirement is the second lowest of all provinces.[48] Furthermore, the Quebec Pension Plan has made no move to increase the age of entitlement.[49] The jurisdiction is an exemplar that within Canada forced retirement can be eliminated without damaging the interests of workers, while also supporting the proposition that a strong labour movement is essential in ensuring such an outcome.

Conclusion

Mandatory retirement reveals some of the cleavages within the Canadian working class. Workers with relatively high-paying jobs and good benefits—such as educators, physicians, and others in professional occupations—are most active in opposing mandatory retirement at a predetermined age. Often workers in these occupations enter the labour force at later ages and may not obtain full-time permanent employment positions until their 30s or later. The leading legal cases reviewed in the previous chapter involve teachers and doctors. For other professional workers, such as judges, mandatory retirement has been set at age seventy-five, while those who are self-employed have no set retirement date. The unions who support mandatory retirement are those that mainly represent blue-collar, service sector, and clerical workers, whose members typically commence their working lives at an early age.

The fact that Canada has not yet eliminated mandatory retirement to the same extent as the United States, Australia, and New Zealand is explained in part by the opposition of the labour movement. In Canada, organized labour has not experienced the retreat that has occurred in the other nations and remains relatively strong. In other words, Anglo-Saxon liberal market economies with low or declining union density, where unions are relatively weak, will see mandatory retirement abolished earlier than nations with higher union density and more influential unions.

Notwithstanding the opposition from unions, the public policy framework (including both legislation and judicial decisions) on mandatory retirement in Canada is likely to shift in the next several years. As outlined by Gillin (Chapter 1), Canada's political and bureaucratic elite has increasingly called into question forced retirement policies. Of course, the specific nature of any changes remains uncertain, as does the long-term impact on workers. In any case, the date of withdrawal from paid employment is certain to remain a controversial issue in the relationship between labour and capital.

Many labour movement leaders, even those who support fixed retirement, expect that human rights legislation will be amended and that greater flexibility in retirement past age sixty-five will become the norm. Although fears that eliminating mandatory retirement will ultimately result in longer working lives and less generous pensions are not unfounded, the outlook for Canadian workers need not be a pessimistic one. The Canadian labour movement remains considerably stronger than its counterparts in the United States, Australia, and New Zealand, and Canada's social security net is substantially more robust, especially with respect to health care, than that of the United States. That Quebec has abolished forced retirement and maintained other aspects of its income security system is testament that protection for workers is not inconsistent with greater freedom in retirement age.

Given the pressures to increase the flexibility of retirement options for the labour force, Canada's unions have the opportunity to play an important part in ensuring that such flexibility is to the advantage of workers. One strategy the labour movement could use would be to allow retired union members to retain full union rights, with the exception of strike votes and collective bargaining agreement votes. In this way, the collective voice of workers could not only be made stronger but also more inclusive, with retirement for individual workers representing less of an abrupt and complete withdrawal from workplace concerns. The failure by organized labour to fully recognize, and respond to, the changing demographic and labour market conditions, and the call for human rights for all workers, will grant greater control to employers and the state to redesign retirement and associated policies.

5

"Fastened to a Dying Animal": Images of Decay, the Social Construction of "Old Age," and Mandatory Retirement

Allan Irving

"Fastened to a dying animal." The phrase is from W.B. Yeats's meditation on old age, "Sailing to Byzantium," and graphically captures our culture's decline-and-decay imagery of older people. Whenever the subject of advocating for the abolition of mandatory retirement in the university comes up for discussion, one guaranteed response is that should this happen, it will require the immediate imposition of performance indicators or measures of competence for those sixty-five and older. Such a response reveals, in all its distressing clarity, just how deeply ageism has burrowed like a tapeworm into our cultural practices. Over lunch recently, a colleague and I began constructing a "doddering index" that could be used by those obsessed with carrying out surveillance on their colleagues to discern any incipient signs of mis-performance: rambling in lectures, making odd and quirky remarks, endless and meaningless historical references (how great it was living through the 1960s), dressing in uncool clothes, outdated eyeglasses, wearing clothes visibly soiled by food spills, mumbling, and talking to oneself. This chapter seeks to suggest that the doddering index—while meant as satire—offers a faithful reflection of attitudes replete in modernist culture.

No less a figure than the celebrated Canadian doctor William Osler—among his other accomplishments, he was a key figure in establishing the medical school at Johns Hopkins University—is remembered for waging an exuberant and nimble battle against what he discerned was the decay of "old age." When he left Baltimore to take up a position at Oxford, he delivered a farewell lecture in 1905 called "The Fixed Period." He lifted the title from Anthony Trollope's 1881 novel, which told the story of a gruesome fantasy society where people over sixty-seven would be chloroformed and painlessly put out of their misery. Osler's interpretation of a fixed period was that the productive years of life lay between twenty-five and forty: "[T]he effective, moving, vitalizing work of the world is done between the ages of twenty-five and forty—these fifteen golden years of plenty, the anabolic or constructive period, in which there is always a balance in the mental bank and the credit is still good."[1] At Johns Hopkins, Osler had been relentless in making it known that in his view medical faculty over the age of forty exhibited signs of decay and decline in productivity, efficiency, and adaptability.[2] Trollope's novel and Osler's pronouncements fit with a late nineteenth and early twentieth century emergent set of beliefs that people should be forced to retire somewhere between ages sixty and sixty-five.

In 1968 a prominent American gerontologist, Robert Butler, created the term ageism to designate a form of discrimination similar to sexism and racism. Ageism is a process of engaging in the systematic stereotyping of, and discrimination against, people simply on the basis of their chronological age. It is the belief forged from particular discourses that people over a certain age (always arbitrarily determined) are somehow less capable, less intelligent, and certainly less deserving of social privileges attaching to younger people, such as the simple right to continue being employed. Ageism is often given a tripartite definition. First there is an affective component, displaying certain negative feelings toward older individuals or groups; second, a cognitive dimension of beliefs and stereotypes of older persons; and third, a behavioral factor of how we actually behave toward older people.[3]

Meanings of age have been produced in modernity as fundamentally revolving around a duality, a binary of young and old that affirms a position of negativity and subordination for those who are older.

Certainly the narrow and limited definition of age (18–65) in many human rights laws has allowed employers to put into practice mandatory retirement (firing without cause), a practice that lays bare the deep roots of ageism in our culture. Professor McKenzie Leiper's personal story in Chapter 11 (fired after only nine years in her job) proffers a revealing account of this cruel policy. Up until recently, retirement and mandatory retirement were not viewed as ageist since retirement was seen in and of itself as a desirable institution. In the post-World War II era, the idea of retirement prevailed over other ways of thinking about older persons.[4] This was marked by a flourishing of income security and pension plans such as the 1951 universal Old Age Security plan in Canada and the Canada and Québec Pension Plans in 1965. There was as well the "aggressive marketing" of retirement as a desirable ideal and as a way to convince those in the labour force that mandatory retirement provided a valuable and constructive alternative to work for older people.[5] A whole retirement industry began to develop with experts increasingly dispensing advice on how to live a worthwhile life outside the workplace. The elderly from the 1950s on became imbricated in a tangle of retirement planning, experts on retirement lifestyles and endless financial advisors. The prevailing tenor of the age is reflected in Robert Keemeier's 1961 collection, *Aging and Leisure: A Research Perspective into the Meaningful Use of Time*:[6]

> [I]ndustry tends to view with suspicion the productive capacities of older workers, particularly when younger and seemingly more efficient persons are available.... The existence of a governmental social security scheme along with private pension programs presumably takes the sting out of retirement, and from management's point of view makes it a desirable alternative to attempts to provide suitable jobs for those eligible for retirement.[7]

Recent psychological studies indicate that older workers remind their younger peers of the imminent threat of death:[8] age cleansing in the workplace may be the result. There are other possible interpretations, though, that follow a more historical, social constructionist storyline. The remainder of this chapter will explore three dimensions that coalesce to form the category old as an essential social construction: (i) the creation of categories and classifications by the Enlightenment/modernity; (ii) the imagery of decay and the abject; and (iii) the discourses of measurement so prevalent in our culture.

The Creation of Categories

In deconstructing the notion of old and its attendant institutional practices like compulsory retirement, part of our task is to undo and recast Enlightenment notions of universals, foundations, and essences, of atemporal and ahistorical "facts." The eighteenth century Enlightenment imparted the central ideas of universals and binaries that have shaped the discourses of modernity,[9] including the field of gerontology. In order to undermine and challenge elitist claims to the authority and traditions of the monarchy and the church, the Enlightenment first had to establish the idea of universal autonomous human reason, a capacity that, as Kant argued, inhered in all persons. Universals such as the discovery of scientific laws and the establishment of universal essential facts through categories and classifications, usually expressed in some binary form such as young/old, were an essential ingredient in fashioning modernity.[10] A central thrust of the Enlightenment was to establish grounds for disinterested, objective, universal truth, old age being one of these foundational truths.

Most, if not all, of the other chapters in this book imply various social constructions of age and retirement. Kerr and Beaujot (Chapter 6) probe the demographic construction of public discourse on the meaning of old age. Pupo and Duffy (Chapter 7) describe the personal reconstructions inherent in growing older and retiring from paid employment. Gillin and Klassen (Chapter 3) analyze the possible changes to the legal construction of mandated retirement; and Gunderson and Hyatt (Chapter 8) and Kesselman (Chapter 9) debate the meaning of different socio-economic constructions of this practice.

There are three specific modes or practices[11] by which human beings are made subjects: classification/categorization practices, dividing practices, and self-subjectification practices. It is these three practices that have created subjects in modernity and the various ordering, controlling, and disciplinary practices that modernity's institutions represent. Geriatrics and gerontology developed in the late nineteenth and early twentieth centuries as human sciences that categorized and classified the aged, and that began the process of constructing knowledge and social practices that fit the established classification. In order to study and further refine the category of old age, expert professionals were necessary to conduct research on, and to proffer advice and recommendations about, the treatment and management of the category.[12] Mandatory retirement gradually materialized as one method for managing the category of the elderly.

From the modernist viewpoint, "unproductive" individuals needed to be divided and subjected to the discipline and control of institutional practices: asylums, hospitals, prisons, and schools.[13] Self-subjectification practices refers to the myriad ways people create themselves as social subjects through what Michel Foucault described as *technologies of the self:* "Techniques that permit individuals to affect, by their own means, a certain number of operations on their own bodies, their own souls, their own thoughts, their own conduct, and this in a manner so as to transform themselves, modify themselves, and attain a certain state of perfection, happiness, purity, supernatural power."[14] In a sense "old age" rather than being an essential, grounded foundation of a certain subjective identity is, at least in part, created by subjects' repeat performances of age learned through the technologies of the self that mandate and inscribe on the body certain ways of being. Thus, categories satirically proposed by the doddering index mentioned above may have some reflexive traction in reality. Mandatory retirement, as well as being an institutional practice, is a technology of the self, a self-subjectification.

Imagery of Decay and the Notion of the Abject

One of the most graphic images of decay in our popular culture is Ridley Scott's 1982 film *Blade Runner,* where the white/bright cities of modernity have been replaced "by a city of darkness, night, chaos and delirium." The final battle scene in the film between the two chief protagonists takes place at the Bradbury Building, designed in 1893 after the technocratic, rationalist notions of Edward Bellamy's 1888 novel *Looking Backward.* In the film, however, decay and deterioration have fully advanced to a state of burst pipes and an eerie hallucinatory space.[15] Here I want to pursue the argument that the Enlightenment's obsession with cleanliness, purity, order, progress, rationality, and certainty resulted in social practices like mandatory retirement to keep unsettling images of decay, the smell of putrescence, and abjection contained in a sharply defined category.

Lurking just below the surface of modernity's relentless pursuit of the new, with its imagery of fluorescence and growth, lies chaos, rot, and filth. In the binary of young/old, the condition of the young is inevitably founded on the sight/site of the waste, blight, detritus, and alterity of old age. In the tenebrous and marginalized world of the "elderly" as it has been imagined so often by modernity, an often fallen and debased world with social horizons narrowed by entropy and the proximity of death and the stillness of bones, Enlightenment optimism can quickly turn to ash. Wishing to preserve an illusion of ontologically secure bodies, it becomes compellingly necessary within the realm of Enlightenment imagery to remove intolerable and distressing signs of decay and abjection from our midst through brutal and cruel practices like mandatory retirement. It seems that there is an Enlightenment imperative to secure images of youth, health, and the pristine and to deny the fragility of these images, their thin illusions vaporizing as decay and disintegration slowly ascend.[16]

The person who introduced the term geriatrics to the medical community in 1909, the Viennese-born American physician Ignatz Leo Nascher, surrounded the burgeoning specialty with essentialized and universal images of bodily and mental decay and abjection. In his famous 1914 text, *Geriatrics: The Diseases of Old Age and Their Treatment*, Nascher described old age as a process of tissue and cell degeneration. The central thrust of *Geriatrics*, and hence its contribution to the construction of gerontological knowledge, portrayed how the elderly were to be classified according to the decay of their bodies, and in chapters on thin hair, brittle nails, dry and loose skin, uneven muscle texture, slackened jaw, loss of teeth, and improper posture he told the sad tale of decline:

> [T]he most prominent mental characteristic in old age is an overwhelming interest in self, a selfishness which gradually subordinates every other interest in life to the welfare of the individual.... [It] is also the cause of his suspiciousness, his egoism and temperamental changes.[17]

Not to let his case rest there, Nascher plunged on in another section of his book to accord the aged universal characteristics:

> When the mind becomes impaired he [the elderly man] neglects his person in every direction until he becomes obnoxious to those around him. He cannot accommodate himself to a progressive order or to modern ideas, he becomes old fashioned, even queer, while those nearest to him try to humor his whims until patience is well-nigh exhausted.... The firm insistence upon hygienic measures for his benefit and welfare, which necessarily impose some exertion on his part, is resented as a hardship and creates a dislike of those who are most interested in his welfare.[18]

Another major historical contribution to the now thick sediment about notions of old age and decay comes from the renowned American psychologist G. Stanley Hall, who, amongst other accomplishments, was president of Clark University. His investigations into, and pronouncements on, the meanings and signs of old age were set out in his 1922 *Senescence: The Last Half of Life,* a book stuffed with decay imagery and metaphors of decline and waning powers and abilities. Hall argued that what he deemed "senescence" started to visibly set in when men were barely into their forties and for women at even younger ages. As he saw it, the human life course, with its perceptible progression to decay meant that men were at their most productive between forty-five and fifty-five. After that there was a rapid and steep decline into what he termed "senectitude."[19] Hall's fertile and inventive mind made ample use of his psychological knowledge to enlighten his readers with a rich tapestry of personal characteristics typical of the aged. His itemization included: lowered ideals and degenerate morals, a growing inability to distinguish right from wrong and true from false, a decline in emotional expressiveness and an accompanying stodginess and cynicism, and a disinclination to obey religious teachings.[20]

Being the dutiful psychologist, Hall urged that his elderly subjects take up inspiring hobbies such as stamp collecting and that everyone over the age of thirty receive psychological counselling to deal with their obsessive thoughts about growing older. His litany of the decaying and abject body was thoroughness itself:

> The old are particularly prone to develop peculiarities … [and] the roster of them on the negative side would be long. There are faults in table manners, modes of eating and drinking … mastication may be noisy or otherwise subtly disagreeable, or there is slobbering, clumsiness, or neglect of common conventionalities once observed. The toilet may be neglected, the attire soiled or spotted or imperfectly put on, and so a look-over is needed before we go out. We … make noises in the presence of others that once we only permitted ourselves when alone.… The voice is impaired in volume, richness of inflection, and articulation; our face or form are no longer aesthetic objects; we mislay things and invoke those about us to help us to find them; and are tediously slow in mind and body. In addition to all this we become pryingly overcurious, fault-finding, exact and forgetful. … On all such matters we should make frequent self-surveys.[21]

Hall was, it appears, far in advance of his time; we have a ready-made doddering index available should compulsory retirement be done away with.

The Enlightenment/modernist master narrative is built around a story of progress, growth, and flourishing that, before we know it, turns into a decline and decay story. Decline and decay have become a fixed foundational truth about aging—"a structure of feeling as deep as this enacts a world, as well as interpreting it, so that we learn it from experience as well as from ideology."[22]

One of the Enlightenment's key progenitors, René Descartes, in establishing a foundation for scientific knowledge, separated the dispassionate and objective mind from the unruly passions and chaos of the body. Descartes wanted to ensure that the mind as it pursued its quest for certain knowledge and essential truth would be unencumbered by the messy, disheveled and often unpleasant abject body in all its corporeal expressions.[23] The notion of abjection refers to what is seen and felt as degraded, vile, unpleasant, unclean—filth, waste, dung—an unwanted "otherness." The abject creates in us a desire to expel, to cast out or away in order to protect our identity boundaries.

> [I]t is not lack of cleanliness or health that causes abjection but what disturbs identity, system, order. What does not respect borders, positions, rules .... Abjection is immoral, sinister, scheming, and shady: a terror that dissembles, a hatred that smiles ... a friend who stabs you .... We may call it a border; an abjection is above all ambiguity. Because while releasing a hold, it does not radically cut off the subject from what threatens it [e.g., images of being old]—on the contrary, abjection acknowledges it to be in perpetual danger."[24]

The category "socially abjected groups" has entered the discourse recently where the case is made that grasping the notion of abjection can provide us with a greater "understanding of a body aesthetic that defines some groups as ugly or fearsome and produces aversive reactions in relation to members of those groups." From this vantage point, "racism, sexism, homophobia, ageism and ableism are partly structured by abjection, an involuntary, unconscious judgement of ugliness and loathing."[25] Anthropology offers additional insights into the construction of abjection and its association with dirt, the unclean. Dirt represents anything that is not in its assigned place and threatens to disrupt order. Using the dirt metaphor, those we wish to mandatorily retire represent symbolic dirt and the disruption of orderly practices.[26] Those who we age by culture and our institutional practices have for far too long been shunted off into spaces of cultural marginality, dirt to be swept away and the neat and the clean restored.

In our workplaces, we want to ensure that material bodies do not put us in danger of having to experience the abject. Mandatory retirement is one technology for disciplining and controlling bodies in the workplace and as a way of expelling bodies before they start expelling matter that is supposed to remain inside the body—for example, "farting, burping, urinating, spitting, dribbling, sneezing, coughing, crying, sweating."[27] We have a desire to avoid leaky, messy bodies, bodies that are literally too fluid and flow out beyond their safe contained selves. Once the body becomes too obvious, as it is suspected will happen to the bodies of those who are older, we are almost forced to confront reason's "underside," its "negative, inverted double."[28] Certainly since Descartes, the mind, especially in the university, is associated with positive characteristics such as "reason, subject, consciousness, interiority, activity and masculinity." The body is another story and has been viewed largely in negative terms such as "passion, object, non-consciousness, exteriority, passivity and femininity."[29] The contributions by MacGregor (Chapter 2), Munro (Chapter 10), and McKenzie Leiper (Chapter 11) all testify to the destructiveness of a regime of imposed retirement on body and self.

From a modernist perspective, the body and its associated framing technologies like mandatory retirement "represent the hard 'facts' of empirical reality, the ultimate justification for positivism and the Enlightenment quest for transcendental reason."[30] Bodies, though, can also be useful sites for a disruption of enlightenment universalist notions of a "natural body." Bodies are socially inscribed and therefore can be reinscribed or redescribed. A universalist approach that sees aging as a part of a natural life course detaches bodies from their concrete locations, a detachment that has allowed us to create bodies through metric means and ready solutions, a materialism that in the case of mandatory retirement can lead to self-redundancy, isolation, and sorrow. The construction of bodies by metric means drops us into the final theme: the discourses of measurement.

Discourses of Measurement

Daily life requires rules and orderly practices if it is to be coherent according to discourses of measurement.[31] Applying this discourse to old age, we could imagine that the field of gerontology as a specific body of knowledge and a technology has played a part through its discourses of measurement, statistical analyses, life-stage imaginaries, and burgeoning definitions in moving old age from a kind of vagueness and something ambiguous to a much more focused and defined entity. Once discourses of measurement, with their retinue of experts and supposed exactness, become the only legitimate way for knowledge to be created, other ways of knowing, such as the "body as a knowing instrument" become obsolete.[32] Discourses of gerontological measurement are intended to gain some measure of control over perceived decay, chaos, breakdown, and instability, and to dispel ambiguity and uncertainty. Decay and instability are seen as a wildness that needs to be tamed and the practice of mandatory retirement—a form of cultural measurement—has been one method for reinstating coherence and order at the point (age 65) of potential chaos.[33]

Conclusion

The Enlightenment and modernity exhibited an almost obsessive desire to yank chaos, however it was defined, back into some sort of orderly system.[34] Perhaps rather than trying to order the chaos of human affairs, we need to find ways to "celebrate momentary clarities" and to inhabit the chaos.[35] The forging of novel ways of seeing things has the potential to lead us away from a kind of totalizing mapping in the construction "old age" that has allowed for authority, arbitrary power, and destructive control. New postmodern ideals of heterogeneity—the abolishing of mandatory retirement and the introduction of individual choice—will assist in collapsing universals that create the very categories that trap us so often in unhappy spaces. Postmodern pluralism enables us to gather new energies for political action and it allows for a return of the subject "in the guise of a politics of new, ignored, and different subjectivities, sexualities, and ethnicities."[36] Age can be reconstituted and recomposed as a constellation of new subjectivities different from those contained in modernity.

In resetting the agenda towards viewing ourselves as multi-faceted subjects, multiple subjects of becoming, we might invoke Foucault's notion of power as a process that flows incessantly between internal and external forces. Its fluidity gives those who oppose mandatory retirement the encouragement to know that where there is power there is also the possibility for resistance. Eliminating coercive practices like forced exit that turn people at sixty-five into reified objects and impoverished beings may offer the individual an opportunity to work on the self almost as an artistic project. "We have to create ourselves as a work of art," advised Foucault. "[C]ouldn't everyone's life become a work of art?"[37] We can shatter the baleful practice of mandatory retirement against the sharp edge of our desire for a rich world of multiple relationships rather than the one we have of frozen simplicity.

6

Demographic Change and Mandatory Retirement in Canada

Don Kerr & Roderic Beaujot

In the early 1980s, Kingsley Davis[1] introduced the idea that population aging could disrupt traditional accommodations across generations. In his discussion of age relations and public policy, future demographic change was understood as potentially leading to a nascent conflict of interest (or class struggle) across generations. Population aging could prompt a society to transfer resources from the young to the old, partly because of the demographic weight of the elderly. In subsequent research, demographers have rarely used such strong language, and have moved away from such exaggerated claims.[2]

Most advanced industrialized countries will experience substantial population aging over the next several decades. Governments fear escalating costs of public pensions for a growing population of older people. Meanwhile, policy-makers worry about the health care demands of large numbers of older people. There are questions, too, about the consequences of reduced labour force participation, slower or even declining rates of labour market growth, and an older workforce. Discussions of these matters often suggest an implicit conflict of interest between the young and the old. The shift in age structure is understood as potentially undermining the standard of living of younger workers, who would ultimately have to contribute to the social security and health care costs of an increasingly top-heavy age structure.

Despite slowing population growth and population aging, Canada, like many countries in the world, continues to permit mandatory retirement. How does the Canadian situation compare with other nations, and how can a public discussion of fundamental demographic trends inform the debate on mandatory retirement? Should Canada expect labour shortages if current trends continue, and what do we know about their potential timing? Further, as mortality and morbidity rates decline, to what extent have Canadians been successful in postponing old age through longevity and better health? Since older Canadians are healthier than ever before, what is currently known of the determinants of retirement, of the characteristics of those who retire early or work beyond the age of sixty-five? These demographic and labour market questions must surely be linked to arguments for abolition or retention of mandatory retirement.

Demography: Declining Natural Increase and High Immigration

On July 1, 2004, Canada's population was estimated at about 31,900,000 inhabitants, up by about 0.9 percent from the previous year.[3] While many might consider this a relatively modest rate of population growth: it is greater than in most other developed nations and is only slightly down from what characterized the last few decades of the twentieth century. For example, the average growth rate of the population over the full period 1971–91 was 1.2 percent annually, dropping slightly to 1.0 percent over the 1991–2001 period.[4] To draw international comparisons, the annual growth rate for all of Western Europe in 2003, taken together, was only 0.24 percent, less than a third of the current Canadian growth rate.[5]

Although Canada's population is growing more rapidly than those of European countries, its pace of population growth is comparable to its neighbour and major trading partner, the United States. Yet fertility in Canada is significantly lower than in the United States—currently 1.5

births per woman in Canada, compared to about 2.1 births per woman in the United States.[6] Yet few countries accept more immigrants per person than Canada. The 2001 Census indicated that over 18.4 percent of Canada's population was foreign-born,[7] which compares to 10.4 percent in the U.S. in 2000.[8] Since 1994, net international migration has contributed more to Canada's overall population growth than has natural increase. Over the 2000–2002 period, net international migration has accounted for about two-thirds of Canada's population growth, which compares with less than half of the population growth in the United States—or more specifically about 45 percent.[9]

Sensationalist publicity on the 2001 Census argued that Canada's population was on the verge of implosion. Canada's national newspaper, the *Globe and Mail*, reported that Canada's population could begin to decline by as early as 2010, an assertion that mystified demographers.[10] Population growth is expected to slow considerably as we move into the twenty-first century, although demographers disagree as to the extent to which this will occur. In effect, births will outnumber deaths for many years to come because of the shape of Canada's age structure. The number of adults at low mortality ages (including the baby boomers) outnumbers by a substantial amount those in their latter years of life, where the risk of death is particularly high. Even though there are fewer births than needed for replacement, the cohorts at reproductive ages have been sufficiently large over the last several decades to ensure more births than deaths. This situation will definitely change as we move further into the twenty-first century.

Using Statistics Canada's last round of population projections, births will likely continue to outnumber deaths until 2025, whereas the overall growth rate will remain positive until at least 2041. According to this medium-growth scenario, Canada's population will increase by about 20 percent over four decades before stabilizing, a much lower rate than the 70 percent growth of the last four decades of the twentieth century. Although demographers do not anticipate an imminent population decline, the pace of population growth will slow considerably. The dynamics of population growth in Canada are very different from elsewhere in the world. In fact, many European countries are experiencing a negative rate of natural increase and are already completely reliant on international migration to maintain population. Germany and Italy register more deaths than births, as do several of the countries of Eastern Europe and of the former Soviet Bloc.[11]

Population Aging in a Context of Declining Growth

Canada's population is aging. Population aging typically takes the form of an increase in the numbers of the old and a fall in the number of children and young persons. This aging of Canada's population is well represented in Figure 1, which presents the population pyramids for 1981, 1991, and 2002. In inspecting these age pyramids, there are a few features that are particularly striking, including (1) the relatively large size of the baby boom cohort (currently in their forties and fifties) and (2) the rather pronounced shrinkage that has recently occurred at the bottom of Canada's age pyramid.

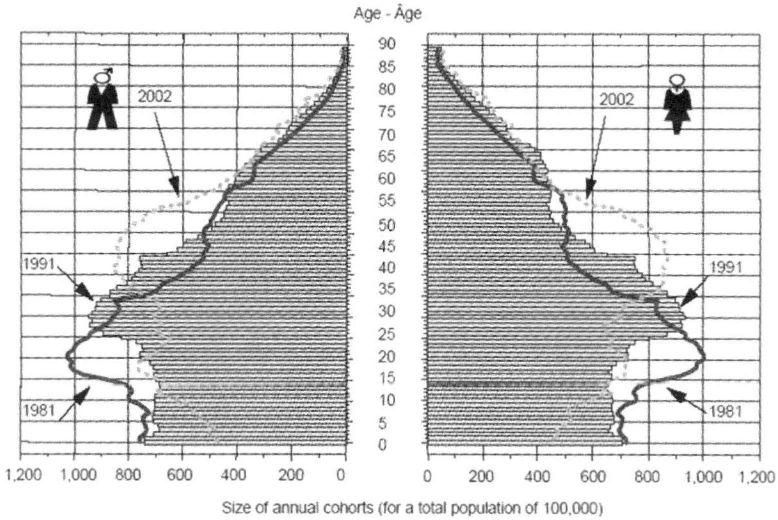

Figure 1. Age Pyramid of the Population of Canada: July 1, 1981, 1991, & 2002
Source: Statistics Canada, 2003. Annual Demographic Statistics. Ottawa: Statistics Canada catalogue no. 91-213.

The analogy of a "pig in a python" has frequently been used by the popular media to describe the passage of the large baby-boom cohort through its life cycle. Due to its mere size, this cohort has had a major impact on several societal institutions. Figure 1 demonstrates that the large cohorts born during the baby boom passed through their young adult years in 1981, were well-established in the labour market and family life by 1991, and in 2002 were moving into and through "middle age." The shrinkage at the bottom of the population pyramid is represented in the 2002 pyramid, in comparing the relative number of teenagers to the number of preschoolers in Canada. This major contraction at the bottom of the age pyramid has received far less attention than the baby boom.

The annual number of births in Canada has been steadily falling for well over a decade. Since 1989, the annual number of births has fallen by almost 20 percent, from about 403,280 in 1989 down to 327,187 in 2002.[12] This is fully 32 percent less than the 479,275 births at the height of the baby boom in 1959. With the passage of time, both these features of the population pyramid (the baby boom, and contraction at the bottom of the population pyramid) will continue to have major ramifications for Canadian society. Consider, for example, that the 1959 cohort will reach age sixty-five around about the same time as the 2002 cohort enters the labour market (in 2024). Canada's population is currently aging due to low fertility and because most of the recent gains in mortality have shown improved conditions at adult and older ages.

It is often thought that immigration will solve the problem of population aging, that is, that young immigrants in the labour force will pay the social security costs of an aging society, especially the costs of health care and pensions. But immigration has little effect on the age structure.[13] Aging will continue regardless of immigration. In fact, the age structure of the foreign-born is older than that of the Canadian-born population. The effect of immigration on the age structure can be appreciated by comparing the median age of immigrants on arrival to that of the Canadian population. The median age of immigrants was relatively stable, averaging 25 years for each year between 1956 and 1976, then increasing to 27 in 1981–86, 28 in 1986–90, and 30 in 1994–99.[14] The median age of the entire Canadian population has changed much more, rising from 26.3 in 1961 to 37.6 in 2001. In effect, the median age of arriving immigrants was one to two years younger than that of the receiving population in 1961, compared to seven to eight years younger by 2000. Both immigrant arrivals and the receiving population have been aging, but arrivals remain younger on average. However, the overall effect is rather small given that immigrant arrivals represent a small part of the total population. Other demographic phenomena, including the movement of the baby boom through the age structure, lower fertility, and mortality reductions at older ages, have a larger effect on the age structure than the arrival of immigrants.

Other measures confirm that immigration has only a small effect on the age structure. For instance, simulating population change after 1951 as a function only of births and deaths produces a 1981 population with an average age that is only 0.5 years older than the actual average observed in that year. Stated differently, the 1951–81 immigration reduced the average age of the 1981 population by half a year.[15] As another example, with no international migration over the period 1951–2001, the median age in 2001 would have been only 0.8 years older than it actually was.[16] Similar results are obtained with projections into the future. The Statistics Canada population projections based on the 1986 census produce a median age in 2036 almost two years younger under high immigration than under zero immigration.[17] This means the arrival of 200,000 immigrants per year for fifty years would reduce the average age by only two years. In comparison, the median age of the Canadian population increased by 2.3 years between 1996 and 2001.

Europe and North America diverge significantly. Population aging is at a more advanced stage in Europe, where Chesnais speaks of an "inversion of the age pyramid."[18] At the turn of the century, the proportion over sixty-five is already over 18 percent in Italy and 16.4 percent in Germany. The figure for Canada (at 13 percent) is slightly lower than the average for the more developed countries, and it is not expected to reach 18 percent until after 2016. Nonetheless, the Canadian change is now as rapid as that of European countries, and it will become more rapid once the larger baby-boom generations move into retirement ages. Whereas in Canada just over one in ten people were over sixty-five in 1986, some fifty years later, in 2036, the medium projection suggests that about a quarter of the population will be over sixty-five. Along with a median age of some forty-five years, and more than 12 percent of the population aged seventy-five and over, this will make for a rather different demographic profile.

Canada's Labour Force into the 21st Century

Just as Canada's overall population is not expected to stop growing, so too the size of its labour force is projected to grow for several years. Yet just as the pace of overall population growth is expected to slow considerably, so too the pace of labour force growth is expected to slow. According to Statistics Canada's most recent projections, the population aged 18–64 will increase at least until 2016, and perhaps for many years beyond this.[19] As the majority of Canadians aged 18–64 are currently employed (that is, about four out of five persons are either employed or looking for work), this rise in population would translate into growth in Canada's workforce.[20] The extent of growth is uncertain, since we cannot know future labour force participation rates nor the manner in which these may vary by age and sex.

Future labour force growth is tied to Canada's immigration policy. But for decades, Canada has not had a settled view of immigration's role in maintaining its labour force. Canada's political climate has shifted from time to time, as politicians, interest groups, and the general public conclude that immigration is either the cause of all of our problems or the solution to much the same problems.

According to the 2002 Annual Immigration Plan,[21] 70 percent of labour force growth in Canada throughout the 1990s was due to immigration. This figure is derived by dividing the number of labour-force participants who arrived as immigrants between two specific points in time relative to the corresponding change in the size of the labour force. Although not likely the intent, this way of describing growth in labour force implies that immigration is the driving force of labour force regeneration, while ignoring the impact of "internal entrants" to the labour force. Statistics Canada has projected that if current trends continue "it is possible that immigration could account for virtually all of the labour force growth by 2011."[22]

Many have the impression that soon the sole source for regenerating Canada's labour force will be immigration. It is highly misleading to say that our labour force will not be renewed unless we have immigration, as this implies that there is little regeneration by all of the people leaving Canadian schools to enter the labour force.

When the labour force is growing relatively slowly, as is now the case, the above statistics are not very meaningful. For instance, if the labour force grew from 10,000,000 to 10,000,001 and only one immigrant moved to Canada, then 100 percent of the labour force growth would be due to the arrival of that one person. This method in estimating the impact of immigration obscures a very dynamic situation underlying labour force growth in Canada. For example, approximately 408,000 people reach age twenty each year, while the annual number of immigrants is about half of that (at 210,000). In this context, an overemphasis on "immigration"—or "external sources of entry"—leads to a neglect of potential "internal sources of entry." If we estimate that 90 percent of those who reach labour-force age will be in the labour force at some point, and that 90 percent of immigrants aged 15–64 will also be in the labour force, then we would have 507,000 additions to the labour force, of which 72 percent would be due to internal recruitment and only 28 percent due to immigration.

By ignoring internal entrants, we implicitly downplay an important source of labour force renewal, that is, the regeneration by more than 400,000 people leaving Canadian schools each year to enter the labour force. It is a distortion of Canada's demographic and economic development to focus almost exclusively on immigration as a source of labour force renewal. Immigration is an important source of recruitment, but there are other ways in which the labour force is renewed.

It could be argued that excessive dependence on immigration would lead to neglect of other sources of recruitment. That is, we may fail to make the necessary investment in education for labour force entrants and/or ongoing labour force training or retraining for older Canadians—with a knock-on effect of neglecting population groups with historically low labour force participation. As about 20 percent of Canadians aged 18–64 are not in the labour force, there is room to increase overall participation rates and thus add to the labour force.

Older Canadians in particular are underrepresented in the labour force. Only about 55 percent of persons aged 55–64 participated in the workforce in 2003.[23] This at least partially follows from the fact that many Canadians choose to retire before their sixty-fifth birthdays. But withdrawal from, or non-involvement in the labour market is not always a matter of choice. Among Canadians aged sixty-five and over, the proportion involved in the labour market plummets, as only about 1 in 14 persons have jobs, and almost half of those who continue to work do so only on a part-time basis.

Figure 2 demonstrates the relatively low participation rates of older Canadians, by presenting age/sex-specific labour force participation rates for the period 1951–2003. Over the last half century, there have been some rather dramatic changes for men and women. Particularly striking is the phenomenal increase in the labour force participation of women, although there continues to be an important discrepancy, particularly at older ages. This increase has been counterbalanced, at least partially, by more modest reductions in labour force participation by men. For example, the participation rate for women aged 25–44 reached an all-time high in 2003, up to about 81 percent. Among women aged 45–54, participation rates have continued to increase through to the present, up to an all-time high level of 79 percent. This substantial rise in female labour force participation has more than offset trends in the opposite direction for men. Participation rates by gender have tended to converge across all age groups. Women made up almost half (or 46 percent) of Canada's workforce in 2003, although they were significantly more likely to be working part-time (with about 1 in 4 for employed women compared to fewer than 1 in 10 for men).

Among older men (aged 55–64), there has been a substantial decline in participation rates, from about 87 percent in 1951 to only 59.1 percent in 1998—rebounding to 65 percent in 2003. Among Canadian men aged 65+, about 40 percent were in the labour force in 1951, dropping to about 11 percent by 2003. Among women aged 55–64, the participation rate has reached an all-time high of 46 percent in 2003, albeit continuing to be much lower than among younger age groups. Among women aged 65+, labour force participation has always been extremely low and has, for the last many years, been at about 5 percent.

Canadians wind down their commitment to paid employment as they move into their latter fifties and into their sixties. Yet these older age groups are now set to grow at a dramatic pace in the next decade. Statistics Canada's medium growth projection suggests that the number of persons aged 50–59 will grow by 42.6 percent between 2001 and 2011, whereas the 60–64 age group will rise by fully 55.8 percent.[24] This is the by-product of the front-end baby boomers moving towards retirement age, widely referred to as the "aging in the middle" of Canada's age structure. Over this same period, only modest growth is projected with respect to Canadians of labour force entrance age.

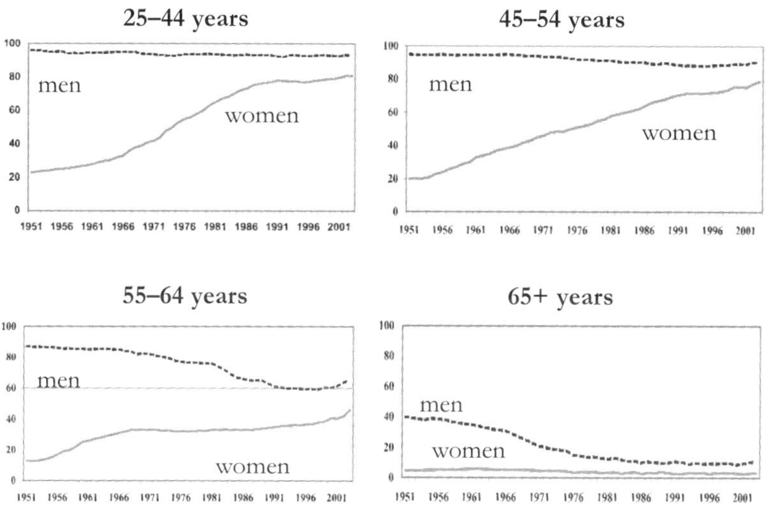

Figure 2. Labour Force Participation Rates by Age Group, Sex and Year, 1951–2003

Sources: Denton, F. and B. Spencer (1998). "Population, Labor Force and Long-Term Economic Growth." *Policy Options* 19:3–9.

Statistics Canada (2004). Labour Force Survey, December 2003. *The Daily*, 10 January, pp. 2–5.

Over the next decade and beyond, Canada, like most other Organisation for Economic Co-operation and Development (OECD) countries, will witness an increasing population share of older persons, who have on average the lowest participation rates of all Canadians. Whether this directly leads to labour shortages is difficult to forecast, although the Chief Actuary of Canada, responsible for managing Canada's pension plan, has recently emphasized that population aging and corresponding labour shortages will likely be the most fundamental challenge to private and public sector planning as we move into the twenty-first century.[25]

Factors Contributing to Early Labour Force Exit

A proportion of all older persons leave the labour market on a voluntary basis, whereas for many others, it is not a matter of choice. Significant numbers decide to retire early in light of cumulated wealth and savings, while others leave paid employment as a result of health problems, disability, or perhaps difficulties in maintaining a job or finding a new one if unemployed. Some older workers are subjected to layoffs and/or other forms of forced exit from the labour market, never successfully to re-establish themselves in the work force. Other workers may retire, begin to draw their pensions, yet for whatever reason continue to work on a part-time basis. A few may retire early, only to decide to return to the labour market after a period of retirement.

As they approach retirement age, many workers enter into an extended transitional stage, characterized by periods of employment (full-time and/or part-time) interspersed with periods of unemployment and re-employment. McKenzie Leiper (Chapter 11) dramatically illustrates the predicament many women face in this regard. Retirement might be better thought of as a process rather than a well-defined event that occurs at a specific point in time.[26] Retirement is not easily defined, which leads to difficulties in providing an explanation of recent trends, let alone providing a forecast of future trends.

Reliance on "self-definition" can potentially provide misleading information, particularly in light of the stigma that often accompanies being laid off or unemployed.[27] While this stigma persists, there is an increasing acceptance of "early retirement" among Canadians, even among those who could potentially continue to work for many years. A certain proportion of older workers may misreport the reason for exiting the labour force and identify themselves as "retired" rather than "involuntarily unemployed."

Older workers are overrepresented among discouraged workers (that is, those who have abandoned their job search) and also are unemployed for longer periods of time.[28] If an older worker loses his or her job, without the skills for easy re-entry into the labour market, might this same worker claim to be retired—in order to preserve self-esteem?

Older workers experience rates of job turnover like those of much younger workers, but in the event of losing a job, are much less likely to be re-employed, with many leaving the labour force entirely.[29] Given considerable job turnover, a significant proportion of all job separations are in fact involuntary (that is, involving a layoff). In following cohorts who reached age sixty-five during the early 1990s, Rowe and Nguyen estimate that only 51 percent of men and 30 percent of women reported "retirement" in leaving their last job. Although "retirement" seems to express a clear "intention" to voluntarily withdraw from the labour market, most of the other reasons listed in the *Labour Force Survey* suggest an involuntary exit from paid work. The high level of job turnover toward career end suggests that barriers or disincentives to re-employment for older workers may very well be an important issue in the future. Older workers may very well want to continue paid employment if provided proper incentives and real opportunities for re-entry.

Evidence on the extent to which workers may prefer to continue at their jobs, yet are forced to retire, is limited and increasingly dated.[30] As estimated by the federal government in the early 1990s, only about one in five men who were subject to mandatory retirement may have preferred to continue working and fewer than one in ten women.[31] Consistent with recent research on retirement intentions, the majority do not plan to work beyond the age of sixty-five,[32] implying that we should not overstate the importance of legislation relating to mandatory retirement. Regardless of what happens in terms of mandatory retirement, there appear to be many powerful incentives and reasons for early labour force exit. An uncertain proportion of workers leave the labour force at a relatively young age as they have accumulated sufficient wealth and pension benefits. An uncertain proportion leaves the labour force against their will, never successfully re-establishing them-selves in paid work. A minority of all workers are forced from their jobs at age sixty-five due to mandatory retirement provisions.

Health and the Older Workforce

With improvements in population health, the risk of morbidity and serious disability has declined in a pronounced manner for older workers over recent decades. The likelihood of heart disease has declined, as have high blood pressure and arthritis, among other chronic conditions that place limitations on activity.[33] Among Canadians reaching their sixty-fifth birthdays, only about 15 percent report a disability affecting the ability to work that could clearly justify the end of a working career.[34] The lower participation rates of older men over recent years do not appear to be a function of poor health or physically demanding work because both health and working conditions have improved. Policymakers should look elsewhere for an explanation—unemployment, technological change, greater pension benefits, age discrimination, and changing attitudes to work and leisure.

Life expectancy at birth has been steadily increasing in Canada—up to 82.2 years for women and 77.1 years for men—more than two years longer than in the United States.[35] "Disability-free" life expectancy has also risen—that is, the number of years on average one could expect to live in good health and without serious disability.[36] Recent improvements in the state of population health serve to highlight the rather arbitrary character of the marker typically associated with the beginning of "old age" or "normal" retirement—age sixty-five. In light of better living conditions, lifestyle, and quality of health care, the average sixty-five year old (or even seventy-year-old) is not as "old" today as thirty years ago. As Denton and Spencer[37] have argued, this marker for old age should be moved upward to the extent that we succeed in retarding what is inevitable, that is, the aging process. This is good news to the extent that we potentially overstate the future costs of old age dependency when we continue to rely on sixty-five as demarcating "old age."

Despite increased longevity, the median age of retirement has declined from about sixty-five years in the 1970s to reach a low of 60.9 years in 1998, before a slight rise to 61.5 years in 2001.[38] This overall decline in the age of retirement and climbing life expectancy mean a longer period of labour force inactivity in relatively good health. As Canadians have longer lives, yet shorter careers, the majority are no longer in the workforce when they reach this somewhat arbitrary "retirement age" of sixty-five. Although sixty-five has become a much less popular age to retire, it remains the most common, as the intensity of retirement peaks at this age.[39] For those who work in organizations with mandatory retirement policies, there is little choice in working beyond sixty-five.

Population Aging, Declining Labour Force Growth, and Mandatory Retirement

The public debate on mandatory retirement should take into account demographic and labour market trends, particularly given that international migration has a negligible impact on population aging. Demography has changed dramatically, modifying necessarily the debate on mandatory retirement. Today is a time of slowing population growth, population aging, slowing labour force growth, and relatively low labour force participation rates among older Canadians.

Labour force growth is expected to slow over the next decade or so, as the sizeable babyboom cohorts move into age groups that have hitherto had relatively low labour force participation, a circumstance different from the labour market conditions in the North American economy several decades back. For example, although Canada's labour market is likely to grow by about 6 percent over the 2001–2011 period, in stark contrast, the labour force grew by a remarkable 35 percent over the 1971–81 period.[40] Demographic circumstances have dramatically shifted over the last several decades, with major ramifications as we move further into the twenty-first century. In turn, the shift in age structure has had and will continue to have an affect on the supply of labour, pay structures, and management strategies. Regardless of what happens in terms of mandatory retirement, this shift in age structure has implications that swamp any impact of change in retirement policy.

Throughout OECD countries, population aging is occurring to such an extent that many economists now forecast that continued trends toward ever-earlier retirement may very well threaten the actuarial soundness of many pension plans.[41] The Secretary General of the OECD[42] has emphasized the importance of a more active role for older people in the economy and society and has recommended a whole series of policy principles to support this goal. Among them is a fight against "ageism" in all forms and, in particular, removal of any discrimination that might constitute a barrier to employment or re-employment. There is little research about why older Canadians are leaving the labour force earlier rather than later, typically well before the age of "mandatory" retirement. We should know more about the linkages between occupation, income, and health, and how these affect the retirement decision,[43] especially as Canada's population ages during rapid social and economic change—in a movement toward the so-called knowledge society.

As the first of the baby boomers turn sixty-five in 2011, it is obviously far from certain how they will behave in terms of labour force participation, particularly as they are better educated and skilled than the cohorts that preceded them. A fundamental question is whether baby boomers' productivity will decline as they move into their fifties and sixties, or the extent and timing of any such decline.[44] In any case, as the baby boom cohorts move toward their retirement they are much healthier then their predecessors. Using information from the 1998–99 National Population Health Survey in Canada, Hogan and Lise[45] estimate the average Canadian male, upon retirement, can expect to be relatively free of serious activity limitation until after seventy-five years of age, whereas the average female at the same age can expect to be free of such limitation until the age of seventy-eight. Since the average Canadian now retires in her or his early sixties, there is scope to lengthen working life to a significant extent. As working conditions have shifted away from physically demanding jobs, a much larger proportion of Canadians over the age of sixty-five could continue to be gainfully employed. Despite a decline in physical skill, we are increasingly successful in delaying this decline. In a time of potential labour shortages, it makes sense to translate gains in population health into a lengthening working life, with further investment in the upgrading and maintenance of skills throughout the life cycle.

Conclusion

The OECD[46] has recommended that governments set out to strengthen work incentives for older workers, improve their employability, and simultaneously invest in the health and care of the older population. Although many older Canadians take advantage of their wealth and pension benefits to retire at a relatively young age, many others involuntarily exit from paid work. To the extent that this is true, there may be considerable interest among older workers to continue with paid employment with proper incentives and real opportunities. Such lures include a work environment that allows and encourages all employees to maintain and update their skills regardless of age. Older workers are presently less likely than younger workers to participate in job training programs, contributing in turn to lower productivity. It is uncertain if this is a calculated choice by older workers moving toward retirement or the result of management priorities.

Mandatory retirement was easier to justify in times of rapid labour force growth and with a relatively young population. This policy should be reconsidered in light of improved population health, slowing labour force growth, and rapid population aging. Rather than encouraging early exits from work, including forced retirement, there should be incentives for older workers to remain in the labour force. Providing older workers with the "right" to work beyond sixty-five may have an impact, on a symbolic level, in the recognition that older Canadians can continue to make an important contribution to society. This recognition might well change the character of public discourse on the meaning of old age, including its associated rights and benefits, as Canadian society adjusts to this new demographic reality. Changes introduced now in our definition of "old age" could have immense consequences over the longer term. Public debate on the advisability of "mandatory retirement" is thus essential.

7

Locating "Mandatory Retirement" in the Midst of Economic and Social Transformations

Norene Pupo & Ann Duffy

The issue of mandatory retirement can only be fully addressed when situated in the midst of the profound social and economic transformations of the past half-century. Since the 1950s the nature of "work" and the "worker" in Canada have been profoundly altered. Dramatic aging of the labour force, unparalleled movement of women (especially wives and mothers) into paid labour, and the increasing presence of immigrants and visible minorities in paid employment all speak to a dramatic reformulation of the typical worker. During the same time period, the loss of industrial and manufacturing sector employment, the whittling down of public sector work, the explosion in service sector employment, the technological revolution, and the compelling growth in non-standard employment have transformed the lived experience of work. The "new economy," rooted in these post-war developments and intensified by economic globalization, means not only a changing face to the Canadian labour market, but also profound and ongoing alterations in the cultural understanding and experience of work as an element in the life course.

This chapter is an exploration into one aspect of these complexly intertwined developments. Specifically, using a series of in-depth interviews with retirees who are currently employed in non-standard work, we explore the growing tendency of Canadian workers to take on post-retirement non-standard employment. Two seemingly disparate social phenomena—the aging of the Canadian labour force (see Kerr and Beaujot, Chapter 6) and the trend toward non-standard[1] (part-time, contract, self-employment, homeworking) employment—are intersecting. For many Canadian workers, retirement, including "early" retirement, is not an end point in their employment history but rather a marker signifying an important transition. Retirees, for a variety of reasons, often return to paid employment. Since non-standard employment is abundant and, in some instances, attractive, many "retirees" are currently employed in part-time, contract, and self-employment arrangements.[2] This emergent pattern is particularly relevant to discussions of mandatory retirement on several grounds. First, it is consistent with a re-examination of the societal meanings attached both to employment and retirement (Irving, Chapter 5). Retirement, in the restricted sense of a complete and final withdrawal from the paid labour force is, in many respects, increasingly anachronistic. Secondly, the proliferation of post-retirement employees strongly suggests mandatory retirement is, for many workers, not desirable and/or not feasible. These workers are voting with their feet. Thirdly, the diversity of experiences in post-retirement employment underscores the importance of addressing the complexity and diversity of the aging workforce and introducing maximum flexibility into any efforts to structure the life course of employment.

The "New Economy" and Non-Standard Employment Among Older Canadians

In the 1970s much public attention was directed to the problem of "leisure." How would Canadians occupy their time once robots and other forms of technology took over from the paid worker? Over time, the scenario narrowed its focus to the plight of younger,[3] affluent retirees who would attain "Freedom 55" and would need to consider how to optimize their enjoyment of the healthy post-retirement decades awaiting them. With the economic upheavals of the early 1980s, the problems of leisure have increasingly dissipated and the leisure problems of early retirement have been supplanted by the economic and personal problems attendant to mandatory or constructed[4] retirement.

In the past several decades, many Canadians have faced the issue of retirement in an uncertain economic context characterized by frequently poor returns on investments, uncertainties over interest rates, continued inflationary pressures caused by increased costs of fuel, electricity and other household fundamentals, and ongoing concerns about the long-term viability of the Canada Pension Plan and other federally- and provincially-funded sources of support for seniors, especially the health care system. The impact of these economic uncertainties is far from uniform. Women, the socially marginalized, members of visible minorities, recent immigrants, and the disabled are likely to be particularly affected, since as they age they are at greater risk of poverty than other Canadians.[5] Put simply, the prospect of mandatory retirement has profoundly different implications, for example, for women facing retirement with limited accrued pension benefits than for workers with generous benefits. However, regardless of status, few workers entirely escape concerns about maintaining quality of life following retirement. Canadian workers of all ages have been beset by deep insecurity[6] rooted in the unrelentingly high rates of unemployment, decreasing "real" incomes over the past two decades, and for many seniors, the difficult financial circumstances they see their middle-aged children experiencing.

Further, in recent decades, numerous workers have not had the option to decide the merits of "early" or mandatory retirement. Many have faced abrupt unemployment due to permanent layoffs in management ranks and/or the transfer of Canadian industrial and manufacturing jobs to less expensive overseas plants. By 1997, fewer than half (48.5 percent) of Canadians aged fifty-five to sixty-four were actually participating in the paid labour force.[7] As discussed in Chapter 12 (Russell), these older workers often face age-based systemic discrimination along with long periods of unemployment, discrimination in the hiring process, and personal difficulties in blending into workplace cultures more suitable to younger workers. In this context, it is not surprising that non-standard work (notably part-time employment) is attractive to many older workers.[8]

Of course, senior workers cannot be approached as a monolith. Seniors' interest in, desire, and opportunity to continue with or re-engage in paid employment varies widely in terms of many variables, including gender, age, social class, health and well-being, marital status, and personal labour force experience. Women economically marginalized by an episodic employment history, by divorce or single parenthood, immigrant status or membership within a visible minority, or by disability will be under considerable economic pressure to continue paid employment. They will also likely confront considerable obstacles. Caught in the midst of a service economy that increasingly generates non-standard employment (often characterized by low wages, few, if any benefits, little job security, and so on) and an ageist society that denigrates older workers, it is likely that they will have few options outside of part-time, contractually limited, and other non-standard work. In short, for marginalized seniors—women, visible minorities, immigrants, the disabled, low-income seniors— part-time, short-term work is often all that is available, especially given that in recent decades job creation has largely resided in the service sector, which is heavy with non-standard jobs.

Seniors who are economically and socially benefited—for example, professional and highly skilled workers (often able-bodied, white male and female professionals)—will be in a better position to successfully maintain full-time employment.[9] Affluent and high profile seniors—politicians, corporate leaders, and executives—are joined by seniors from a wide variety of occupational groups, including many high-ranking occupations—judges, ministers, professors, lawyers, physicians, auditors, and accountants and senior managers.[10] It is noteworthy, for example, that many of the corporate chieftains in Canada are aged sixty-five or older. William Carroll's recent examination of the corporate elite reported a continuing "greying of the elite," with three-fifths of the elite in 1976 and 1996 being older (aged 51 to 65) and a substantial increase (up from 2.9 percent in 1976 to 6.8 percent in 1996) in the numbers who are post-retirement (age sixty-five).[11] Among this advantaged population, "'post-retirement'" employment, including part-time employment, may be approached less in terms of financial need and more as a source of social connectedness, personal fulfilment, and social standing.

Post-retirement employment, of course, is not solely the preserve of the well-educated and affluent. Between 1996 and 2001 the numbers of men over age sixty-five employed as commercial truck drivers increased by 83 percent. During this same time period, the numbers of senior women truck drivers increased by a staggering 500 percent. A recent issue of *Truck News* commented, "Like a division of Energizer bunnies marching in formation, thousands of senior citizen truck drivers are refusing to slow down and hang up their keys."[12]

Given the various pushes and pulls on the aging workforce, it is not surprising to find an extensive part-time senior labour force. Currently, about one in five public and private sector workers over age fifty-five is employed part-time. Predictably, given the prevalence of women in part-time work throughout their life cycle, the senior part-time workforce is significantly gendered. Women, who as young wives and mothers have participated in part-time employment so as to manage the lioness's share of child-rearing and household work, continue to follow this employment strategy as they age. Reflecting a variety of factors, including out-of-date or limited educational credentials, occupational segregation, an interrupted employment history, and correspondingly inadequate pension benefits and/or increased caregiving responsibilities for aging family members, older women workers are more likely than men to participate in part-time work.[13] In 2003, almost 34 percent of employed senior women worked part-time (compared with just over 27 percent of women younger than fifty-five).[14] In contrast, amongst employed men, 14.6 percent of seniors (55+) and 10.4 percent of men below age fifty-five worked part-time in 2003. The gender patterns are distinctive, with women much more likely to work part-time throughout the life course and men to concentrate in part-time work in the 18–24 and over 55 age groups.[15]

Many men and women seek part-time employment as they age. In doing so, they are buffeted by economic, familial, and personal pressures. Part-time employment may be undertaken as a means to reduce the paid workload; in order, for example, to accommodate personal illness or disability or to care for others, including ailing spouses. Conversely, part-time employment may not be seen as a step back from full-time employment but rather as a step into the paid labour force, bringing with it economic advantages along with possible advances in terms of personal fulfilment and personal connectedness. However, as noted above, participation in post-retirement employment, part-time and full-time, is patterned, notably by gender.

In all occupational categories and industries in both the public and private sectors,[16] women over age fifty-five are far more likely to be working part-time than men over fifty-five, with particularly dramatic differences for those working in retail sales, wholesaling, insurance, and clerical and lower level administrative services. In retail trade, for example, 17.5 percent of senior men compared to 44.2 percent of senior women worked part-time in 2003. Similar differences among senior part-timers occurred in transportation and warehousing (15.5 percent of 55+ men, compared to 37.4 percent of 55+ women), management and administration (24.3 percent of men and 46.6 percent of women), educational services (18.8 percent of men, 36 percent of women), and accommodation and food services (15.5 percent of men and 35.1 percent of women). Similarly, amongst the self-employed, senior women are much more likely to work part-time (45.4 percent) in contrast to comparable males (21.5 percent).

Despite this tendency for women to dominate the ranks of senior part-time workers, more men than women in the 55+ ranks stated they preferred part-time over full-time work.[17] Further, their motivations are different: far more women than men noted personal and family reasons for working part-time.[18] These differences, of course, reflect a lifetime of gendered social and economic realities. By age fifty-five more men than women have amassed private and workplace pension benefits, permitting greater confidence in their move to part-time work, and granting a measure of freedom in their work-time options. They may be limited, however, by commitments to their dependents as well as by a number of lifestyle choices. Women's work histories and work-related decisions follow completely different paths and are constrained throughout their life courses in myriad ways by responsibilities and demands of their families and households. Along with the remunerative aspects of employment, the meaning of work, the social connections derived from work, and the personal satisfactions are socially constructed and guided by gender.

As discussed in detail below, the post-retirement non-standard employment of men and women casts an important light on issues of retirement. It is apparent that many workers want and need paid employment after retirement. Some categories of workers, women and the socially marginalized, for example, are likely to be particularly motivated by economic need. Much as there is an ongoing need for work after retirement, there is also a desire for a rethinking and, possibly, a reformulation of work. Many men, for example, seek out part-time employment to allow for an improved balance in their private and public lives. In short, the process of "retirement" may imperil some workers, especially the most vulnerable, while at the same time providing others with an important opportunity to reorient their working lives.

Post-Retirement Non-Standard Work: Some Observations from a Case Study

In order to bring some lived experience to the discussion of post-retirement employment, we consider the experience of a sample of 18 individuals (12 males and 6 females) who after retirement re-joined the labour force to work part-time. The research is based on in-depth interviews. The interviewees are "retired" in the sense that they have formally retired from their main occupation and are now receiving pensions. Many took early retirement while others retired at age sixty-five. Their part-time work is extremely variable, with some working three days a week, and others working full-time for six months and taking six months off. Some are part-time only on occasion, depending upon the demands of work. Many are also self-employed or contract workers.

The research population was constructed through posted notices and snowball sampling[19] in the Niagara peninsula and in the Ottawa area. There are no claims that this population is representative of post-retirement public sector workers. Indeed, the research group is better educated and more affluent than the "typical" senior Canadian worker, and as such, these workers are very positive about their post-retirement work, often describing it as a refreshingly new direction for them. Like Singh and Verma,[20] who examined "bridge employment" undertaken by retired telecommunications workers, we find workers who are more in transition than facing an abrupt shift in their work/non-work lives.[21]

Along with being advantaged workers, most of our interviewees are male and we report primarily on the males' experiences because of the similarity in their pre-retirement careers and in their backgrounds.[22] However, the women's interviews do provide us with a valuable commentary on the ways in which the final pre-retirement years may evolve into post-retirement employment. For both the male and female respondents, we are provided with a very strong sense of why some workers would actively seek to continue to be employed. Most importantly, this qualitative material reveals the nuances and complex feelings that are attached to employment, both pre- and post-retirement. The research results relate to the issue of mandatory retirement on a number of levels. On the simplest level, the interviews speak pointedly to the inappropriateness of strict mandatory retirement policies. These men and women once again provide clear evidence of the continued productivity and creativity of senior workers (see also Leiper, Chapter 11).

Next, the results suggest that the process of "retirement" may be worth maintaining if it is extricated from its mandatory confines. The opportunity to relocate oneself in terms of work may be invaluable to workers and to their employers. Even those individuals who continue to pursue work that is markedly similar to their pre-retirement work—for example, consulting to the same government bodies that once employed them—speak of valuing the new patterns of work. Many speak with great enthusiasm about the shift in the locus of control. They feel they have much more say in the way their work is scheduled, allowing them to maintain their interconnections with family and leisure activities.

Viewed in more detail, the process of retirement and, in particular, the availability of a base (pension) income have in almost every instance been the catalyst for a significant rethinking and reorientation in interviewees' work and personal lives. For those who continue to work on similar tasks pre- and post-retirement, the ability to rely on a pension income has, for example, injected a greater sense of flexibility into their working lives and this, in itself, has often improved the work experience. Others, for whom retirement has permitted them to escape an increasingly tedious and unrewarding work experience and whose pension income allows them to pursue employment dreams (such as turning a hobby into a source of income), the retirement process has been experienced as a very positive and valued step towards not only autonomy but also personal fulfilment.

Finally, the retirement process appears to trigger in many of these individuals a breaking down of the compartmentalization of their lives—work, family, hobbies, leisure, volunteer activities, community involvement—and permitted these commonly separate elements to blend together in a more organic and satisfying manner. These themes are explored in the following discussion of the research results, followed by an examination of the implications for mandatory retirement.

The Process of Moving to Post-Retirement Employment

Not surprisingly, many of the respondents indicate they were thinking about their post-retirement employment prior to their formal retirement from full-time work. Most were concretely making plans, taking actions to establish the foundation for this new stage of work. However, as documented elsewhere in the research literature,[23] a hiatus often occurred before a particular employment path took hold and was pursued. This hiatus sometimes lasted several months or years after employment and reflects a period of work inactivity or of work experimentation. Interviewees, for example, talk of making personal contacts, taking courses, and seeking advice as they consider their next life-course step and weigh their alternatives. This process is, in itself, conditioned by the post-retirement position of the workers. Their pension income provides them with a reliable, socially sanctioned source of income for this period of transition.

Why Return to Paid Employment After Retirement?

Income Generation

Concern over increasing or maintaining income levels post-retirement is a complex issue. Once again, social class may be a factor. Research suggests that employees from higher income brackets may experience a steeper drop in income after retirement than those whose incomes are lower.[24] Ironically, it would seem that wealthier Canadians may be faced with more momentous and numerous lifestyle adjustments when they retire. In this context, it is not surprising that continued employment is an economically attractive alternative.

Some respondents indicate they wish to maintain their pre-retirement lifestyles and, as a result, income is an important consideration in their post-retirement paid work. However, the need for income is often articulated as part of a complex mosaic of concerns. One individual currently working as a consultant to the Canadian International Development Agency comments:

> No, I really want to supplement my income for the next ten years or so. So that I can have a good retirement pot. And I like to keep busy, and I certainly want to develop my skills and learn things. I always like learning new things. So going to Indonesia and learning about a new country is right up my alley. I just love it.

An ex-assistant deputy minister who is now lecturing at a local university articulates a complex balancing of factors as well: "I do it because I'm retired. And I really enjoy it and I really enjoy the money. It's very nice, but the job is very satisfying. I wouldn't do it for free, but I wouldn't do it if I disliked it." Another individual who has parlayed a hobby into an income-producing venture similarly embeds the economic motive into the pursuit of satisfying work:

> That's one that's missing. Because although it hasn't come out in the interview, I was getting less and less interested in government in terms of what I was doing and when the opportunity came for early retirement I just jumped at it. Because I basically wasn't happy in the way government programs were unfolding and the

> type of role that I was playing ... I was not a very happy camper as such. There was very little job satisfaction in the final years in my work for the government. So my objective with this particular business that I have is two-fold. I wanted to do something I really enjoyed doing and secondly, I wanted to make up the difference between what I was earning and what I'm getting on pension.

In short, while income production is a central concern, many try to construct post-retirement work so that it is challenging, holds their interest, and allows time for family and other personal activities.

Further, any reference to the desire to produce income and/or maintain a lifestyle is in itself a multi-faceted concept since it typically does not refer to the fundamental ability to provide sufficient income to house and feed the individual and his or her immediate family. The motivations implied in economic considerations may range from extra funds for more travel to intense concern about the future economic well-being of the respondent or the immediate family. An ex-career army employee comments, for example, "I retired from the Department of Defence with a pension but the pension is … just enough to meet basic life requirements, so that any sort of lifestyle demanded that I make … additional income." Another individual appears to echo these words: "And we have to pay our expenses, so the money is primary. The thing is to have enough cash flow to have the lifestyle I want." However, later in this interview, it becomes apparent that *lifestyle* incorporates a strongly felt sense of concern and obligation to a son who is developmentally delayed. He continues:

> I want to make sure that there's sufficient cash flow and funds available that will carry him for the rest of his life in a style that he would ordinarily never afford nor be able to have. It's just not going to be possible. And so the money has to be there to fund that for him. And it's going to take a lot. And then if it's not there he's going to have to live on whatever the government provides, or whatever else someone's going to provide, or the little bit of money that might come from insurances. It's going to be gobbled up … in no time, so I don't want him to live that kind of existence.

Another individual, despite his retirement status, indicated he had several young children at home and therefore economic considerations included support for his family. This concern about the future of government-provided income security benefits is echoed by a number of respondents. Many interviewees share the perception, reinforced through the media, that societal support systems (e.g., medicare and social security) cannot be relied upon in the long run—that the burgeoning ranks of retirees or inflation will place an insupportable burden on income-security plans. For those retirees, this rationale for post-retirement employment allows for sufficient resources in their old age, despite any future failures in the support systems.

The Work Ethic

Not surprisingly, many respondents indicated they wanted "to keep busy." An individual self-employed as a consultant on disabled seniors comments:

> Yeah … I didn't retire very well…. I mean I've got lots of hobbies and … I was involved in those hobbies, but all of a sudden I was treating the hobbies like work with deadlines and schedules and all of that. I said, well, if I'm going to do that, I might as well get paid for it. Then my hobbies can go back to being fun. So, that's basically what happened. The need to keep busy kept me, brought me back into the work force.

Other interviewees echo these sentiments: "Why do I work? Partly, I think the basic reason is I'm interested. I don't have any intention to *sit around and do nothing* [emphasis added]. Keeping busy. Number two I like the extra income, which is nice. Three it keeps me in circles of people." Another respondent comments, "… I always have to be busy. It doesn't matter. Something. Something. You know what I mean." Again, the words are echoed by others: "Well, the benefit is it keeps us from becoming lazy and … being like couch potatoes," and "I don't have any intention to sit around and do nothing," and "Keeping busy, yeah, I'm not one to sit around. I mean I would have to do something."

Perhaps a seventy-three-year-old retired chemist currently employed by the Canadian Executive Service Organization and involved in extensive foreign travel sums up the attitude of many: "Well, you know, I say that I want to live until I die. … I don't want to die till I die. I want to do things and it's fun. I enjoy it or I guess I wouldn't do it." In many of the interviews, a sense of moral obligation to keep busy is complexly interwoven with other considerations such as pleasure in productivity and social involvement.

Learning or Maintaining Skills

As noted above, prying apart the various motivators for post-retirement employment is difficult, as one intricately relates to another and few stand alone. The opportunity to explore new skills, particularly when combined with income generation and satisfying work, is an important consideration. One interviewee elaborates:

> I like to grow my work skills as well, which this position allows me to do on an almost daily basis because the community that I'm dealing with is absolutely different from Department of Defence or from National Defence. I mean, I'm dealing on a day-to-day basis with either persons with disabilities or people within agencies that provide support to persons with disabilities. So it's a completely different environment for me.

However, as this respondent goes on to explain, these learning opportunities are understood as contextualized by a variety of other key factors:

> What I recommend [is] that anybody who isn't really satisfied with retirement find some avenue to go out and fulfill whatever they need to be fulfilled … And certainly this [being an independent contractor] is a good way of doing it. I'm certainly busy. Certainly learning. Certainly involved in the community. And earning a little bit of money to keep body and soul together, but maintaining a fair degree of independence on any given day. As I said before, I am required to produce certain things for [social

> agency] and as long as I produce them on time, on budget, nobody bothers me. Nobody tells me how to do it, they just tell me what they want. Learning new skills is complexly interwoven with other perceived positive attributes of post-retirement work—notably social connectedness, autonomy, control over the work process, and economic reward.

Social Connectedness

Although some variability existed in the amount of social contact pursued and enjoyed by post-retirement workers, some respondents indicated that the opportunity to work with others and maintain a social network was an important consideration.[25] As one respondent comments: "Yeah, I like the social environment of work … I don't like working on my own in my little office at home. I mean I can do it, and I'm set up to do it and … I do it when I have to do it, but I'd much rather be in an environment where there's other people around." However, once again, this factor does not stand alone. A computer analyst, retired from employment with the City of St. Catharines who returned as a contract worker, explains:

> So basically it's friendship. I enjoy working there. I enjoy working with the people. I always did. The money, of course, the money is important. Always is. That's very good, I might say. And, I think, keeping busy. Yeah, I can keep busy anyway. In fact, this is interfering with some of my other lifestyle choices…. It's basically—I feel, I feel needed and didn't particularly, especially when the city said we're offering this [retirement package] to get rid of you, effectively.

These remarks, of course, reflect on the larger societal context. As suburban communities have increasingly become deserted enclaves during the day, personal life, friendships and a sense of belonging have steadily shifted into the workplace. Although there is considerable variation in this effect, numerous social analysts are arguing that more and more workers are finding their subjectivity, intimate life, and sense of meaning are strongly, if not exclusively, situated in their work experiences.[26] Viewed from this perspective, retirement may be seen as not only a loss of economic resources, but also a loss of personal connectedness.[27] For many, this social marginalization may not be resolved through participation in golf clubs, travel, and volunteer activity. Indeed, many of the mainstays of community connection—the Lions Club, the Kinsmen, the Canadian Legion—are increasingly viewed as out-of-date and old-fashioned or have been transformed into an extension of the paid workplace—a place to network.

The problem of social connection in the post-retirement period is unlikely to be soon resolved. Long-term trends such as urbanization, internal migration, decreased family size, increased divorce rates, postponement of marriage and child-bearing, and the paid employment of wives and mothers all tend to reduce both the availability of seniors to their children's families as well as the need for seniors in these families.[28] It is not surprising that social marginalization and decreased self-esteem[29] are significant issues for seniors and that post-retirement employment is an attractive alternative for many.

Personal Gratification and Enhancing Self-Esteem

As noted above, the retirement process contains a complex and intense emotional dimension.[30] In a society structured around paid employment and a market economy, the loss of a job and a reduction in income are often experienced as a personally painful double blow. Conversely, the process of continuing to work may be particularly gratifying for older workers who are in a position to share their wisdom, experience, and skills with a younger generation of workers. As discussed by MacGregor in Chapter 2 and elsewhere,[31] there are many individuals whose involvement in work activities is central to a sense of personal satisfaction and fulfillment.

Several of our respondents discuss this aspect. One respondent clearly enunciated the painful implications of the retirement process and the motivation to address self-esteem issues. He said that returning to paid work as an independent contractor

> … helped my self-esteem…. [W]hen you first retire, unless you're really old and tired, the one thing you lose to some extent

is self-esteem. Because all of a sudden you are not what you were before. When I was with the city I was a senior systems analyst and I was responsible for my business and so on. I came home here and that morning, that first morning, what are you?! I'm retired. You know? I wasn't saying independent business owner, which I am, but I wasn't saying that. And yeah, and there was a kind of loss. There was a kind of hollow there, an emptiness kind of thing. It took me a long time before I realized what it was actually, and then, then hey, I am not somebody. I'm not a, you know, in the federal government. They call themselves CF7 or something like that. I'm a CF7. You're supposed to know, oh, that means you're that level of pay scale. So I lost that. And I got it back by activating other things, by doing things like community work and so on and so forth. And canvassing for charity and that kind of stuff. And then getting active in my business, so on. The first, probably the first four or five months I was rattling around feeling kind of worthless, I guess. I'm not saying but feeling it, and not realizing what it was, and later on I said, now I know what that was all about.

Rethinking Work: The Benefits of the Retirement and Pension Process

The evidence from the interviews presented here clearly suggests that the process of retirement is frequently very valuable for employees, providing them with an opportunity to radically assess their life course. One individual who had been an economist in the federal government found in retirement an opportunity to leave what had become a boring job. "Oh … I wish I could have done it earlier. Like, like fifteen years earlier than I did, because I became rather bored with the government. But I stayed with the government until financially, since I was going to get a pension." Retirement provided him with a release and a chance to involve himself in more rewarding activities.

While the process of arriving at post-retirement work may vary considerably, many of these elements—the much-valued sense of personal control in the work process, the excitement or fulfilment of tackling different tasks, and the perceived crucial role of pension backing—appear throughout the interviews. Another individual discussed his sense of increased personal control and satisfaction: "I'm doing what I really have always wanted to do, quite frankly." Interestingly, once again, the transformation includes a new relationship with his wife: "Well, my wife is a partner in the company, partly as my business manager linked to, we're trying to computerize the inventory at the moment." Finally, once again, the pension is seen as a crucial backdrop to this work alternative:

> ... I doubt whether this business would generate the income that I would require to have the same lifestyle as I had when I worked for the public service.... Yeah, I've got financial security in terms of the pension....

Surprisingly, workers who pursue very similar pre- and post-retirement work activities also speak of many of the same elements. A self-employed consultant, whose pre- and post-retirement work was "the same thing," speaks enthusiastically of his sense of independence and control: "[independence on the job]—wonderful. Absolutely wonderful. If I don't like it I can walk away, and say, thank you. I can turn jobs down, so ... it gives me a lot of independence." Once again, the post-retirement arrangement means a new relationship with his wife: "The only other person in my own little firm is my wife, who manages the business for us, in terms of money and bills, and income tax and all that kind of thing." Once again, the pension income is seen as a crucial ingredient.

Finally, it is important to highlight some of the more embedded themes here. First, the retirement process often allows for a dramatic shift in interpersonal/family relations. It seems noteworthy that some of the men make a point of now officially incorporating their wives in their working lives as "business partners." In one instance, a retiree with young (early teens) children in the home, had used his more flexible post-retirement work situation to assume more domestic responsibilities:

> Since my wife has a job [in] which she experiences a fair amount of pressure, which means that sometimes she might have to

> work, it's a high-tech.... Sometimes she might have to work on weekends. Sometimes she might have to work late. So I look after things at home. I get supper. Look after the kids when they come home. See that they get to their music or to their soccer, and we've got two taking music and one in soccer.

Similarly, some of the female retirees indicate that retirement, despite paid employment and other activities, has allowed them more time to develop relationships with children and grandchildren.

The Implications of Non-Standard Post-Retirement Employment for Mandatory Retirement Policy

Canadians are clearly not of one mind when it comes to the issue of retirement and the opinions of older Canadians may be particularly varied. A recent survey by Decima Research found that over a seven-year period, from 1996 to 2003, the number of Canadians opposed to mandatory retirement increased from 20 percent to 33 percent. Not surprisingly, as workers approach retirement age, there is a marked increase in opposition to mandatory retirement with 24 percent of those 18 to 29 years old opposed compared to 39 percent of those over age 50. However, it is also clear that many Canadians anticipate profound change in their participation in paid labour. In 2003, more than one-third of Canadians surveyed indicated they expected to retire before age 60 and 26 percent suggested they intended to retire sometime after age sixty-five.[32]

These results, along with the results of the research presented here suggest there is an enormous and growing diversity in Canadians' approach to employment in their fifties, sixties, and beyond. Structured by important societal dimensions—gender, social class, visible minority status, and so on—some Canadians will be under economic pressure to continue to participate in paid labour. Given good health, manageable domestic and familial responsibilities and reasonable employment opportunities, they may be inclined to defer any withdrawal from paid employment. Other Canadians, as evidenced by many of our respondents, may be strongly influenced by a complex array of monetary and non-monetary factors. As a result of a complex calculation, they may decide to withdraw (retire) for the short or long term, or they may opt for alternative working arrangements (part-time, at-home, contract, and so on). Others may simply wish to continue on with their lifelong careers (MacGregor, Chapter 2).

With these diverse patterns in mind, it would seem that only an extremely flexible approach to retirement is capable of accommodating the complex interplay of factors in individual Canadians' lives. This flexibility might include options for both employees who wish to retire "early" and those who seek to continue working past age sixty-five. More importantly, workers must be provided, through their working lives, with the information and space to carefully evaluate both their personal situations and the broad range of alternatives. For example, alternative working structures such as part-time positions, contract/consultancy positions, paid or unpaid leave, and alternative work arrangements such as telecommuting, should be considered. In this way, employee and employer alike can benefit from ongoing efforts to optimize the fit between worker, work, and life.[33]

Mandatory retirement policy was established very much in a different era, with a kind of "one size fits all" philosophy—a philosophy that at that time and in that social and historical context was at least more meaningful than it is today. As documented throughout this book, the notion that at sixty-five or earlier all or most workers need or want to withdraw from paid labour is simply fallacious, and should be cast aside in favour of an "age-neutral" discourse around retiring or withdrawing from paid employment consistent with a life-course approach to work, family, and time.[34] As echoed in the words of our respondents, many seniors are entirely capable of contributing in a paid work setting and are eager to do so. This capability is often mobilized in the interests of perceived needs—monetary, societal (community service), moral (productivity), personal (fulfillment), and social (connectedness). Further, rather than needing to give way to the next generation of workers, many senior workers are needed to fill highly skilled jobs as well as to provide mentoring and training to younger workers in the field.

But challenges to mandatory retirement do not imply that the rights of workers to withdraw from or reduce their paid workload should be ignored. These rights are particularly important in light of the systemic patterns of inequality embedded in current employment arrangements. For example, while women may wish to adopt a part-time solution to their hectic mid-life schedule of home and children and paid work, their resultant financial situation may necessitate the possibility of continuing paid employment in their senior years. Where a job requires onerous physical labour (for example, firefighting) that is likely to result in physical disability with increasing age, it is sensible to consider regular evaluations of the worker's abilities and desires to perform the required work, as well as opportunities to withdraw and reorient at a relatively young age (for example, after ten years of service). Acceptance (with these kinds of stipulations) that seniors can, if they so desire, continue their paid employment as long as they satisfy the requirements of their employers appears sensible. Particularly for certain categories of workers who have been historically marginalized in the labour market—women, immigrants, visible minorities—the opportunity of continued employment provides a means to make up some of the ground they may have lost in terms of income, accrued pension benefits, and advancements.

Finally, the rejection of mandatory retirement should not, however, negate the clearly significant benefits of what is now termed the retirement process. Evidently, the process of moving toward and through retirement provides an opportunity for intense self-examination and possible reorientation, which are invaluable to both the individual retiree as well as to the maintenance of a vibrant and innovative workforce. Retirement may, as we see from the interviews described here, be an opportunity for personal, familial, and occupational renewal. Future workplace policy needs to maintain this resource. Employees (and employers) should be encouraged to consider ways to facilitate growth and change in their lives. Phased-in retirement, as described above, more flexible and reduced[35] working schedules for workers seeking a better balance in their work/non-work lives, increased opportunities for retraining and education for older workers, and paid and unpaid leave provisions (sabbaticals) would retain many of the benefits of the current retirement process and, in some instances, may enhance the process by making it available to a wider age range of workers. Since pension income is key to these periods of re-examination and reorientation, such a flexible approach to work in the life course would necessitate some ongoing setting aside of income for future sabbatical or retirement periods.

Framed most broadly, the current intersection between the aging Canadian (and international) workforce and the debate on mandatory retirement provides a crucial juncture for opening up discussion about the nature and structure of work. In recent years, there has been considerable examination of the relationship between paid work and family life. With the movement of wives and mothers into paid employment, the traditional work/family split has been eviscerated and, as a society, we are struggling to create new ways both to work and to raise children. More flexible employment options, home-working, child care, and family leave provisions all speak to the efforts to rethink the ways in which work and family figure in our daily lives. The debate around aging and retirement is a similar catalyst for rethinking the complex interconnections not only between work and family, but between work, community service, volunteerism, and leisure. The lives of post-retirement workers suggest that the interplay between work, leisure, volunteerism, and pleasure are complex and nuanced. How we seek to arrive at a fair and equitable arrangement that allows for meaningful work while also permitting and rewarding service to the community and facilitating periods of leisure, renewal, and reassessment is a project that emerges very naturally from any exploration of policies to replace mandatory retirement.

8

Mandatory Retirement: Not as Simple as It Seems

Morley Gunderson & Douglas Hyatt

Mandatory retirement is often associated with such expressions as forced retirement and age discrimination. It is often even confused with notions that there are government policies that compel individuals to leave the labour market at a certain age, like sixty-five. This perception, for example, appears in the Ontario government's Speech from the Throne on April 30, 2003, announcing the government's intention to ban mandatory retirement: "[The government] will also introduce legislation to allow more seniors to remain active in the workforce—retiring at a time of their own choosing, not an arbitrarily *government appointed time.*" [emphasis added].[1] This perception of government policy prohibiting people from remaining in the labour force is further illustrated by such statements appearing in such media outlets as the *Globe and Mail* as: "CPP [Canada Pension Plan] should be more flexible so that it *allows* people to stay in the work force past sixty-five if they want to and are able." [emphasis added, highlighting the implication that the CPP does not allow people to work past sixty-five].[2] It is perceptions and statements such as these, as well as the rhetoric surrounding the debate, that have led to statements like: "The debate over whether to ban mandatory retirement is one of the most misunderstood discussions in the area of labour and social policy."[3] Hence, also, the title of this chapter: "not as simple as it seems."

The facts are that there are no government-appointed times nor government policies that prohibit individuals from working past age sixty-five or any age. Government laws and policies may influence retirement decisions and they may allow (or disallow) private parties from entering into arrangements that involve mandatory retirement, but they do not involve a government-appointed time or requirement to retire.

The issue is further confused by the general failure to make a distinction between being in favour of mandatory retirement and being in favour of *allowing* individuals or their unions to enter into contractual arrangements that involve mandatory retirement. It is distinctly possible to be *opposed* to mandatory retirement at your particular workplace, but to be in favour of *allowing* individuals or their unions to enter into contractual arrangements that involve mandatory retirement. That is analogous to being individually against abortion, but pro choice. Allowing mandatory retirement is the "pro-choice" option in that it allows people to enter into such arrangements.

Banning mandatory retirement, in contrast, prohibits private parties from entering into such contractual arrangements that involve mandatory retirement. In that vein, a truth-in-advertising dictum would replace the phrase "banning mandatory retirement" (which gives the favourable impression of prohibiting age discrimination) with the phrase "prohibiting parties from agreeing to policies that may involve mandatory retirement" (which is effectively what the ban does). That is, banning mandatory retirement involves banning private contractual arrangements—and the conditions under which this should occur should be the focus of the policy debate. This highlights that the relevant policy question is: "*Under what circumstances should private parties be prohibited from entering into contractual arrangements like mandatory retirement that may constrain their decision making in the future, in return for some other presumed benefits that may be associated with the policy?*" This is a far cry from the way the debate is often rhetorically framed as: "*Should governments be allowed to require people to retire at some arbitrary age?*" or, "*Should employers be allowed to engage in age discrimination?*" The rhetorical questions portray a stereotype of those in favour of allowing mandatory retirement as supporting age discrimination, or the notion that older persons are less productive or that people have to be forced to retire for other social reasons.

Focusing the debate around the issue of where private contracting should be prohibited highlights the relevant issues that should surround the mandatory retirement debate, and it highlights that there can be legitimate disagreement about those circumstances. There are times when the state prohibits such contractual arrangements, such as with prostitution (in most countries) or for private contracts of indentured service when individuals agree to work for a sponsor for a fixed period of time in return for having their passage paid to the "New World." There are other times when the state sanctions private contracting even though part of the contract can inhibit our flexibility and freedom. This is the case with marriage contracts and with mortgages and bank loans that we must pay back in the future. Are mandatory retirement agreements more like prostitution or indentured service, or like marriage contracts and mortgages?

The purpose of this chapter is to try to narrow the range of disagreement around that debate and to make the case as to why and when contractual arrangements involving mandatory retirement should be allowed. The prevalence and characteristics associated with mandatory retirement are first discussed, followed by a critical assessment of the conventional arguments for banning mandatory retirement. The arguments in favour of allowing private parties to enter into arrangements that involve mandatory retirement are then advanced. The chapter concludes with practical and specific policy recommendations that try to strike a balance between allowing private contracting, but only when certain safeguards are in place.

Prevalence and Characteristics of Mandatory Retirement

Mandatory retirement is a provision—as part of a company personnel policy, collective agreement, and/or pension plan—that terminates the existing employment relationship at a specific age. In some circumstances (sometimes termed compulsory retirement), the person can be hired back by mutual agreement under a new contractual arrangement. In other circumstances (sometimes termed automatic retirement), the person cannot be hired back. In all circumstances, the individual is not required to retire from the labour force, but may seek employment elsewhere.

Although direct evidence is not available, indirect evidence suggests that about half of the workforce is in jobs that involve mandatory retirement.[4] The typical age of mandatory retirement is sixty-five, the age at which private employer-sponsored pensions as well as public pensions like the Canada Pension Plan "normally" come available. Working in jobs that involve mandatory retirement does not mean that workers are, or will be, involuntarily constrained by the provision in that they would like to work longer once they arrive at that age. Some may quit or leave prior (including through early retirement), some may die, some may continue working with the same or another employer, and some may prefer that retirement date, especially when it is associated with receipt of private and public pensions. Of persons who are retired, approximately 12 to 20 percent report they retired because of a mandatory retirement policy. Of those who retired because of mandatory retirement, about 6 to 20 percent did so involuntarily in that they would like to have continued working. If 50 percent of the workforce works in jobs with mandatory retirement, and 12 to 20 percent of those retire because of mandatory retirement, and 6 to 20 percent of those would like to carry on working, this suggests that one-third of 1 percent (i.e., 0.5 x 0.12 x 0.06) to 2 percent (0.5 x 0.20 x 0.20) of the workforce is involuntarily constrained by mandatory retirement.[5] These are very rough calculations, but they illustrate that the numbers who are involuntarily constrained by mandatory retirement is likely to be very small, and there appears to be reasonable agreement on this point.

These numbers could overstate as well as understate the numbers who likely would carry on working if mandatory retirement were banned. They could *overstate* the numbers who would continue working because some who say they would like to continue working may be making that statement based on the existing contractual arrangements, which could involve deferred compensation, whereby they are overpaid relative to their productivity when older in return for having been underpaid when younger, and they may be subject to little monitoring and evaluation when approaching the mandatory retirement age (these possibilities are discussed subsequently). In such circumstances, individuals may well say they would like to continue working. But if mandatory retirement were banned and these practices dissipated (again discussed subsequently) then fewer individuals may want to continue working. Working in the other direction, the small number who indicate they would like to continue working may *understate* the numbers who, for various reasons, would actually continue working if mandatory retirement were banned. They may have geared their retirement plans to the existing mandatory retirement date, but that may change if there is a paradigm change with no mandatory retirement; they may not be able to afford to retire if public and private pensions and deferred compensation dissipate as a result of banning mandatory retirement (discussed subsequently), and impending skill shortages as the baby boomers retire may make postponed retirement more attractive.

The likely small number who would continue working if mandatory retirement were banned is a two-edged sword in terms of arguments for and against banning mandatory retirement. It could suggest that the costs to employers or to youths who may have fewer job opportunities are likely to be small, since few individuals would carry on working in any case.[6] It also means, however, that any benefits to those who would like to continue working are small, since few are constrained by the policy in the first place. Furthermore, any benefits in the form of reduced labour shortages or savings in public and private pension payouts from postponed retirement are also likely to be small. Is the glass half full or half empty?

Importantly, jobs subject to mandatory retirement tend to be "good jobs" with higher wages, long-term stable employment, pension and retirement plans including generous early retirement provisions, and with the protection of a collective agreement or a formal personnel policy. The so-called "bad jobs," with the opposite characteristics, tend not to have mandatory retirement; they tend to have no formal personnel practices and can be characterized more by the "employment at will" principle, meaning, of course, the employer's will. The fact that mandatory retirement tends to prevail in "good jobs" is an important consideration in the debate, since it highlights that mandatory retirement should not be regarded as an oppressive policy imposed by mean-spirited employers on disadvantaged employees with no individual or collective bargaining power. Banning mandatory retirement, for example, means banning the right of unions to negotiate a mandatory retirement provision as part of a pension plan for their members.

In the next section, the arguments for banning mandatory retirement are advanced and assessed. The assessment is critical, since the argument of this chapter is that the case for prohibiting such private contracting in this area is not persuasive. In the absence of solid reasons for overriding such contractual arrangements, this chapter contends that they should be allowed, perhaps with safeguards to ensure that they are agreements that are freely entered into.

Arguments in Favour of Banning Mandatory Retirement

Age Discrimination and Violation of Human Rights

The strongest argument in favour of banning mandatory retirement tends to be that it constitutes age discrimination and violates the individual's human rights since the person is required to retire (more accurately, their existing contractual arrangement is terminated) based on *age* and not on grounds such as competency. The appearance of age discrimination is fostered by the fact that although all jurisdictions in Canada have human rights codes that prohibit discrimination on the basis of age, four jurisdictions have an age limit of sixty-five in their code. This appears to say that you cannot discriminate on the basis of age—unless the person is over sixty-five! While the age cap exists to except mandatory retirement as constituting age discrimination, it does create a loophole in that persons over age sixty-five do not have the normal protection of the human rights codes. The legal issues are explained more fully in Chapter 3 by Gillin and Klassen.

While age discrimination undoubtedly exists (and unfortunately may be tolerated more than other forms of discrimination[7]), labelling mandatory retirement as constituting age discrimination appears inappropriate. As indicated previously, the mandatory retirement age is part of a contractual arrangement between employers and employees or their union. Employees agree to accept this mandatory retirement age in return for other benefits such as pensions, deferred compensation and reduced monitoring and evaluation. There is flexibility in the age of retirement; it does not have to be sixty-five—in fact, it does not have to exist if the parties do not agree to it. The University of Calgary, for example, does not have a mandatory retirement age, while the Universities of Alberta, Athabaska, and Lethbridge have it (all being in a province that allows mandatory retirement provided it is accompanied by a bona fide retirement plan). Similarly, while Manitoba exempts universities from the ban on mandatory retirement, the University of Brandon has no mandatory retirement age, while the University of Manitoba and the University of Winnipeg have negotiated mandatory retirement after age sixty-nine with their unions.[8] Some may label those universities and their unions that have negotiated mandatory retirement as discriminating against older workers. An alternative perspective, however, is that they face different constraints and concerns, and that the private parties themselves can negotiate over these matters with mandatory retirement emerging in some situations and not in others—as is the case in the workforce in general. Furthermore, other contractual arrangements involve age and they are not automatically labelled as discriminatory. This is the case with compulsory school laws or child labour laws or the CPP, which is available at age sixty-five.

With respect to the possibility that mandatory retirement violates individuals' human rights by requiring them to retire at a pre-specified age, it is important to keep the perspective that individuals may rationally enter into agreements that inhibit their freedom and flexibility in the future in return for other benefits. This is the case, for example, for marriage contracts, student loans, and mortgages. As well, if the benefits associated with mandatory retirement are such elements as pensions, higher compensation, and reduced monitoring and evaluation, then it is not obvious that an individual's human rights are violated when their agreed age of retirement comes to fruition. If banning mandatory retirement leads to more monitoring and evaluation, and ultimately possible dismissals, the human rights and dignity of those individuals would also be jeopardized.

Age can also be regarded as different from other enumerated grounds in human rights codes such as sex or visible minority status in that all individuals can expect to reach age sixty-five (with good fortune), but all persons will not be female or of a visible minority. If we agree to a contractual arrangement that involves a future commitment to retire at a specified age, and this is regarded as discriminatory since it is based on age, then presumably we are discriminating against *ourselves*—or at least ourselves in the future. This is different than discriminating against well-defined *other* groups, whose oppression we may benefit from. In essence, mandatory retirement involves individuals entering into an *inter-temporal* agreement in which they *themselves* agree to a rule that may constrain them at a later age. Since these individuals ultimately are applying the rule to themselves and not to others, the result should not be regarded as age discrimination.

Poorly Informed or Irrational Agents

This leads to a second argument that is advanced in favour of banning mandatory retirement. That is, such private contracting should be prohibited because the individuals may be poorly informed about the arrangement, and even if they are reasonably informed they may behave irrationally.[9] This is especially the case for younger people who may not be informed about the pension and retirement plans of their jobs. Even if informed, they may simply discount the future too heavily and behave myopically, caring only about their present circumstances. When they are older and mandatory retirement is impending, they may feel that they made a mistake and regret that they entered into such an arrangement. This is possible even though they may have benefited from the policy earlier (for example, if they received their jobs and possible promotions because others were retiring and vacating jobs) and they may even be benefiting from any deferred compensation, reduced monitoring and evaluation, and the pension protection that tends to accompany mandatory retirement.

While such scenarios are possible, it is difficult at best and hazardous at worst to ban such private contracting on the grounds that individuals may be poorly informed or irrational, except in the most egregious of circumstances. Suicide is illegal in large part for these reasons and especially because the decision is irreversible. But other private contractual arrangements that involve individuals making inter-temporal decisions they could regret in the future are not only allowed but are sanctioned by the state. This is the case with marriage contracts, student loans, and mortgages, all of which yield benefits but also involve commitments into the future. Mistakes are obviously made, but we do not "protect people from themselves" by prohibiting such contractual arrangements, nor do we ban youths from becoming police or firefighters because they may underestimate future risks. Mistakes can be made both ways—that is, by not buying a house or not getting married—just as they may be made by not accepting a job because it involves mandatory retirement and then regretting having bypassed a good job with pension protection and other desirable features. Surely, the answer is to provide as much information as possible on the commitments and future implications that are involved, but then to allow such commitments among "consenting adults." This is especially the case since jobs that involve mandatory retirement, as discussed previously, tend to be "good jobs" with the protection of a formal personnel policy or collective agreement and accompanied by the financial security of a pension plan and other retirement benefits.

Some may question, as do Kesselman (Chapter 9) and Munro (Chapter 10), the notion that mandatory retirement as part of a collective agreement represents the wishes of all individuals in the bargaining unit, as opposed to the wishes of some interest groups or even simply the union leaders. But this is a potential issue for any provisions in the collective agreement. Some individuals may not want a medical plan (perhaps because their spouse has one), others may not want a pension plan (perhaps because they would prefer higher current wages), others may not want family-friendly policies (perhaps because they have no families). Yet we do not ban unions from negotiating these provisions simply because some members may not want them. The collective bargaining process involves articulating the internal union trade-offs amongst the members and coming up with a package that benefits most members, an issue explored by Klassen with Forgione (Chapter 4). Furthermore, those issues have to confront the equally compelling issues of management, such as costs, competitiveness, and managerial prerogatives. The preferences (and constraints) of employees and employers may change over time such that one or both parties may want certain provisions bargained away (including mandatory retirement). The virtue of leaving it up to the private parties to do so (rather than governments prohibiting certain provisions) is that the various trade-offs, both within employee groups and between employees and employers, have to be considered in that process. If there is concern that union leaders may not adequately represent the interests of certain groups, those groups have recourse through the "duty of fair representation" requirement for unions to represent their members. These concerns arise in any collective actions that involve democratically elected leaders and collective decision making, including ones made by governments. Some individuals may resent defence expenditures; others may resent social expenditures. But the nature of a democracy is that these preferences have to be traded off against each other, with not all individuals being happy with the outcome of each component.

Saving on Social and Other Expenditures and Dealing with Labour Shortages

A further argument advanced for banning mandatory retirement is that it will lead to savings in age-related social and other programs if people continue working. As well, people who continue working because of a ban on mandatory retirement can reduce impending labour shortages associated with the retiring baby-boom population. There are, however, problems with these arguments.

First, they are somewhat inconsistent with the argument also advanced by proponents of banning mandatory retirement that a ban will have no negative implications since few people will want to continue working in the first place. If they do not continue working, then they cannot also yield savings in social and other expenditures or reduce impending labour shortages.

Second, for there to be a saving in age-related social and other programs, then the benefit payouts for such programs must be reduced if people continue working. But that is not the case, for example, with the CPP in Canada, where recipients receive their full benefits at age sixty-five whether or not they continue working. (In the United States, Social Security payouts are reduced if people continue working.) In Canada, there would be no savings unless clawbacks are introduced or the age of entitlement is raised (both of which have happened in the United States,[10] where mandatory retirement is banned). If reducing age-related benefits is the intent, then this should be explicitly stated as part of the debate since this is the very concern of the trade union movement and others: banning mandatory retirement is starting down a slippery slope where benefits to seniors will be whittled away on the ground that they can keep working. A recent statement based in the context of banning mandatory retirement would suggest that such possibilities are on the policy radar screen in Canada: "'The federal government should re-consider automatically handing out Canada Pension Plan benefits to Canadians at age sixty-five,' Bank of Canada Governor David Dodge said yesterday."[11]

The same concerns apply to employer-sponsored pension plans—that is, unions suggest that employers will feel less pressure to provide pensions and other retirement benefits if workers can continue working indefinitely. If there is to be a cost saving to employers, then pension payouts must be reduced perhaps by making no actuarial increase to the pensions of those who postpone retirement so that there will be a penalty to postponed retirement. If that is the case, then employees are penalized by such pension reductions. If the reductions are not made, then there is no relief to employers on their pension obligations. One cannot have it both ways. The issue is further complicated by the fact that those jurisdictions that have banned mandatory retirement in Canada (i.e., Quebec, Manitoba, and the federal government for its employees) have also banned such pension penalties.[12]

The issue is not that raising the age of the CPP entitlement or reducing or clawing back benefits is a wrong policy, or that pension penalties to balance postponed retirement in employer plans are wrong. These are legitimately debatable issues, but they do not seem to be well understood or articulated in the debate over banning mandatory retirement.[13] If the intent is to reduce such benefit payouts, then that goal should be explicitly stated, as should the mechanisms whereby any reduction is to occur. Reductions need not occur automatically, and if they do occur, they have implications for the retirement security of seniors.

Third, if mandatory retirement is a barrier to reducing upcoming labour shortages, then the practice should dissipate on its own accord or be bargained away by employers. There is nothing stopping employers from abandoning the practice if it is part of their personnel policy, or from bargaining with their unions over its removal if it is included in collective agreements. If the union wants the policy retained, then employers may have to "give something up" in return for its removal—for example, employers may have to guarantee that there will be no pension penalties for those who carry on working. But this is a desirable outcome, since it means that the benefits and costs of the practice have to be considered by the parties closest to the issues at their workplace. It is not simply unilaterally removed by government fiat.

Reduce Poverty Amongst the Aged

Banning mandatory retirement is sometimes advocated as a step in the reduction of poverty amongst the aged. The logic at first seems simple. If some older people are required to retire, this reduction in their earnings may push them into poverty. As stated in the Report of the Special Senate Committee on Retirement Age Policies, "Mandatory retirement often amounts to a sentence of continuing poverty."[14] Or as stated by Lynn McDonald, "If mandatory retirement is sustained … the work option will evaporate and disadvantaged older workers will be forced into poverty."[15]

Such an argument, however, ignores the fact that mandatory retirement is associated with good, well-paying jobs with pensions and other retirement benefits. Those who retire from jobs with mandatory retirement are very unlikely to be in poverty, unless there is some unusual circumstance that knocks them into poverty, and if that happens it is very unlikely to be mitigated by any ban on mandatory retirement. In fact, if the banning of mandatory retirement leads to a reduction in private and public pension income, this is *more likely* to lead to poverty among the aged. Poverty among the aged is much more likely to occur among people who did not have a job around their retirement age or who were part of the working poor and not subject to mandatory retirement. It would be a rare occurrence for someone to be in poverty because they had to retire due to mandatory retirement; the opposite is more likely to be the case.

Disparate Impact on Groups Like Women and Immigrants

A final argument that is often advanced in favour of banning mandatory retirement is that it tends to have a disparate impact on groups, like women and immigrants, who have not had sufficient time in the labour force to accumulate income and the service credits for pensions, and who may have relatively low wages.[16] Empirical evidence, for Canada, for example, indicates that women tend to receive lower pension benefits than do men for a number of reasons: they are less likely to work in jobs with pensions; when they do, they are less likely to accumulate service credits that enhance pension benefits and especially that make them eligible for subsidized early retirement programs, and their lower wages get compounded into lower benefit accruals.[17] If subject to mandatory retirement, women and immigrants may not accumulate the service credits that enable them to retire with substantial pension income or to receive the generous early-retirement pensions that have minimum service requirements.

This can occur. But our CPP benefits are also based on years of contribution to the fund, and if these or other groups have short periods of contribution, they will receive lower benefits. These issues are known in advance and are part and parcel of the decision to immigrate or to leave the labour force. Women who leave the labour force may be penalized in various ways—loss of seniority, experience, labour market networks, and a possible deterioration in human capital. It is not obvious that mandatory retirement should be singled out for being banned because it may inhibit women and immigrants from accumulating service credits. The magnitude of any effect may also be very small given that women's labour force participation is now approaching that of men. Most importantly, however, women and immigrants may benefit from the job and promotion opportunities that are created by persons who retire and vacate jobs. To the extent that the older workforce reflects the older demographic profile of a male-dominated white workforce, then such renewal may facilitate achieving employment equity objectives through any new job and promotion opportunities.

A wide range of arguments have been advanced in favour of banning mandatory retirement. They are often contradictory, poorly understood, based on questionable premises, run the risk that they harm the very persons they are designed to help, and are seldom based on empirical evidence. While it may be too strong a statement to say that they should be rejected, they certainly merit deeper scrutiny.

Arguments in Favour of Allowing Mandatory Retirement

As indicated previously, the argument in favour of *allowing* mandatory retirement is essentially in favour of allowing private parties to enter into contractual arrangements, especially in circumstances where they have reasonable individual or collective bargaining power, and there are other benefits associated with the arrangement. The previous discussion of the prevalence and characteristics associated with mandatory retirement highlighted that mandatory retirement meets these conditions, since it is generally associated with "good jobs," pensions and other retirement benefits, and the protection of a collective agreement or formal personnel policy. In such circumstances, the argument presented here is that the private parties should be allowed to enter into such arrangements even though the benefits and costs may differ between employers and employees and across different employees. If, for example, employers prefer mandatory retirement and employees would prefer to have no mandatory retirement, then employers would have to compensate employees for such an "adverse" work arrangement just as they would have to compensate them for shiftwork or a fast-paced work environment. The compensation would likely take forms that would be complementary to mandatory retirement, such as through pensions, early retirement options, and retirement benefits. Unions do not likely bargain *for* mandatory retirement because they want their members to retire at age sixty-five (employees always have the option of leaving), but rather because of the retirement package that is associated with mandatory retirement.

A virtue of allowing the private parties to work out those arrangements is that they can vary according to the perceived costs and benefits of those who have to live with the arrangements. As those costs and benefits change, so will the arrangements. For example, if employers find that mandatory retirement inhibits them from filling labour shortages, they can change their policy from automatic retirement to compulsory retirement, so that they can retain some employees, or they can raise the mandatory retirement age or abandon the practice altogether. If they negotiate with a union over these issues, they may have to ensure other elements as part of this change, such as ensuring there are no pension penalties for those who postpone retirement. In such circumstances, the practice will change according to the perceived costs and benefits that the parties themselves associate with the practice. It will not change because governments deem it inappropriate under any circumstances or because they are hoping that abolishing mandatory retirement will save on age-related social expenditures.

What, then, are the benefits that the private parties themselves see in mandatory retirement[18] that fosters it being prevalent in "good jobs" and as part of collective agreements or formal personnel practices?

Facilitate Renewal and Job and Promotion Opportunities

Retirements can facilitate employee renewal and the job and promotion opportunities for others, including youths and employment-equity groups. This can be particularly important to expand the opportunities for target groups for whom such job and promotion opportunities are otherwise restricted.

It is certainly the case that it is inappropriate to "get rid of" some groups simply to facilitate renewal[19] and create job and promotion opportunities for others, even if the "others" may be target groups for whom society would like to create more jobs and opportunities. Firing without just-cause would also facilitate renewal and open job and promotion opportunities, yet we do not sanction (indeed we prohibit) such firing. But mandatory retirement, with its associated retirement benefits and the fact that it applies to all persons covered by the practice in that workplace, is not like firing without just cause.

It is also the case that the labour market is dynamic and that a job occupied by an older person does not mean that other groups cannot obtain jobs. That notion is subject to the lump-of-labour fallacy that assumes there are a fixed number of jobs in the economy.[20] While the assumption of a fixed number of jobs is a fallacy for the economy as a whole, it is not a fallacy for particular work environments for which there are a relatively fixed number of jobs. In universities and other organizations where there are a relatively fixed number of positions, new slots are often predicated on retirements. In fact, the positions are often labelled as the "so-and-so" slot named for the person who is retiring.

In essence, facilitating renewal and job and promotion opportunities is relevant in environments where such renewal is important. Again, this argues in favour of allowing the private parties to work out such arrangements, since the costs and benefits of renewal and job and promotion opportunities differ across such environments.

Facilitate Planning

Having a well-defined retirement age also facilitates planning on the part of both employers and employees. For employers, it assists in succession planning, since they have a reasonable prediction of the numbers who are leaving and when they are leaving, as well as the characteristics of those who are leaving. It facilitates their costing of age-related benefits, such as pensions and health and disability plans. It avoids the complications that can arise, for example, when long-term disability benefits cannot be renewed after age seventy.[21]

Of course, if mandatory retirement were an egregious act, it would not be justified simply because it facilitates planning for employers. Nevertheless, the fact that it is a mutually agreed-upon arrangement suggests that the planning benefits to employers can be considered as one of the rationales for mandatory retirement.

Mandatory retirement also facilitates planning on the part of employees, since the termination date is known in advance and with a degree of certainty. It will not come as a "surprise," and therefore, will foster planning for retirement, saving for retirement, and preparing for possible transitions into retirement.

Reduce Monitoring and Evaluation and Facilitate Retirement with Dignity

Mandatory retirement reduces the need for constant monitoring and evaluation of older workers, since any deviations from acceptable performance are more likely to be tolerated given the fixed retirement date. The same applies to work teams if an older worker's contribution to the team falls off. This risk of deviation from acceptable performance is especially likely if the variance of performance is higher for older workers, perhaps because of illness.

In contrast, if mandatory retirement is banned,[22] employers will need to monitor and evaluate older workers more carefully, in part to prepare for the possibility of an unjust dismissal claim if the worker has to be dismissed— and such dismissals will inevitably occur.[23] This, in turn, will put more pressure on older workers, since the threat of dismissal will be more prominent as will the monitoring and evaluation of their performance. When workers "retire," it will not be obvious that they retired voluntarily or were dismissed or were pressured to leave. In contrast, with mandatory retirement they are more likely to "retire with dignity" at the normal retirement age and with the security of a pension.

Of course, monitoring and evaluation is a normal part of the employment relationship for all workers. Nevertheless, it will become more stringent if mandatory retirement is banned and employers have to dismiss some older workers rather than "waiting it out" until they retire. This has to be kept in mind when assessing the pros and cons of banning mandatory retirement.

Facilitate Deferred Compensation

Mandatory retirement facilitates deferred or back-loaded compensation in longer term employment relationships whereby individuals are under paid relative to their productivity when they are younger and overpaid relative to their productivity when they are older.[24] Mandatory retirement can provide a termination date to such a contractual arrangement; otherwise, it cannot exist since the overpayment period could go on indefinitely.[25] A termination date is a necessary equilibrium condition to equate the expected present value of the overpayment and underpayment periods necessary for such arrangements to exist. Deferred compensation[26] can occur in the form of pension benefit accruals or pension rights that employees accrue as they accumulate service credits and seniority-based wage increases in their latter years.[27]

Deferred compensation, in turn, can be advantageous for both employers and employees. For employers, it deters unwanted turnover and hence fosters the firm's willingness to train its workforce, since it is more likely that firms can recoup their quasi-fixed training costs if employees have an incentive to remain with the firm to get their deferred compensation.[28] It fosters loyalty and commitment for the same reasons and encourages employees to have an interest in the financial solvency of their employers—perhaps even encouraging concession bargaining if it ensures receipt of their deferred compensation (e.g., pension rights). It discourages employee "shirking," since they risk dismissal and hence loss of their deferred compensation (i.e., deferred compensation is analogous to a performance bond held by the firm). It discourages "lemons," or poor performers, from applying to the firm, since their productivity will be revealed over time. It enables periodic and retrospective monitoring (based on past performance) with good performers retained and paid their deferred wage.

Employees can also prefer deferred compensation[29] for a number of reasons. It facilitates an ever-increasing consumption profile and saving for retirement. It enables periodic, retrospective monitoring, which also may be preferred by employees. It can provide a higher lifetime wage profile if it has the positive productivity-enhancing effects discussed previously.

Clearly, there are a variety of reasons for employers and employees to prefer deferred compensation. Mandatory retirement provides a termination date to such contractual arrangements and therefore facilitates its existence.

The existence of such deferred compensation may also shed light on why it is perfectly rational for individuals to enter into arrangements that involve deferred compensation along with mandatory retirement, and then prefer the banning of mandatory retirement when they approach that age. If mandatory retirement is banned, they will likely continue to receive a deferred wage for a period of time until such compensation profiles are adjusted downwards. In the meantime, however, they will receive a "windfall" gain from the abolishment of the mandatory retirement part of that contractual arrangement. This can also explain some of the voting pressures in favour of having governments abolish the mandatory retirement part of the contractual arrangement now that the large baby-boom population is approaching the typical mandatory retirement age.

There is a wide range of benefits associated with mandatory retirement for both employers and employees. Furthermore, those benefits and costs may vary across different workplaces (for both employers and employees) as well as over time. As such, the main rationale for mandatory retirement is that it tends to be a contractual arrangement that is mutually entered into by both employers and employees for a wide range of reasons. Governments should prohibit such mutually beneficial arrangements only under egregious circumstances. The analysis of this chapter suggests that mandatory retirement does not fall into that category for reasons outlined.

Appropriate Policy Response

The legislative initiatives with respect to mandatory retirement in Canada are quite complex.[30] Conventionally, mandatory retirement has been allowed because the human rights codes of the various jurisdictions have an age cap of sixty-five, beyond which the code does not apply. While this allows mandatory retirement in the sense that it cannot be contested as discriminatory, it does so at the expense of not providing conventional protection against age discrimination beyond age sixty-five. Four jurisdictions—British Columbia, Saskatchewan, Ontario, and Newfoundland—have such age caps. All other jurisdictions except Quebec and Manitoba have removed the age cap (which would appear to ban mandatory retirement by allowing it to be contested as constituting age discrimination), but they have exempted bona fide retirement or pension plans (which effectively allows mandatory retirement because it almost invariably exists as part of such plans). Only Manitoba has effectively banned mandatory retirement through its human rights code by not having an age cap *and* not exempting retirement or pension plans. Quebec has also banned mandatory retirement through its labour standards legislation, although the certainty of this can be questioned, given a court decision that allowed people to be terminated at the age of sixty-five on the grounds that only those up until the age of sixty-five were guaranteed employment.[31] The federal government also does not have mandatory retirement for its civil servants,[32] but this is an option that can be followed by any employer; it is not because federal legislation has banned mandatory retirement. In essence, all jurisdictions in Canada (except for Manitoba and Quebec) effectively allow mandatory retirement, either by having an age cap in their human rights codes or by exempting bona fide retirement or pension plans.

The legality of mandatory retirement where it is allowed has been confirmed in the well-known "trilogy of mandatory retirement cases" by the Supreme Court in 1990, where mandatory retirement was deemed to be discriminatory but demonstrably justified in a free and democratic society because its social benefits (emanating from the reasons discussed previously) exceeded its social costs, including the possible infringements on the equality rights of some older workers. Nevertheless, subsequent legal decisions have qualified the conditions under which mandatory retirement can exist. A British Columbia Supreme Court decision,[33] for example, overturned a mandatory retirement requirement on the grounds that it had not been effectively communicated to the employee and the employee had not explicitly accepted it as a condition of employment. A British Columbia arbitration board decision, upheld by a court of appeal,[34] indicated that the Supreme Court decisions did not imply that mandatory retirement was *always* to be allowed. Rather, it was to be allowed when its social benefits exceeded its social costs and that could be determined on a case-by-case basis.

As well, in all jurisdictions, mandatory retirement at a specific age can be required if it is a bona fide occupational requirement (BFOR) of the job. The requirements for this are very stringent; hence, it tends to be limited to situations that involve public safety, such as with airline pilots, police, and firefighters.[35]

While there is considerable debate over whether mandatory retirement should be allowed, there is considerable agreement[36] that the institutional and legal barriers that inhibit older persons from continuing in employment should be removed (or at least re-evaluated) where possible. Such barriers can include age limits like sixty-five in some human rights codes that thereby do not provide protection against age discrimination for persons over sixty-five,[37] pension penalties to postponed retirement, clawbacks in the Old Age Security (OAS) system and in Guaranteed Income Supplements (GIS), the requirement to "substantially cease working" to be in receipt of early CPP benefits, penalties for those who delay receipt of CPP until age seventy, and especially after age seventy, requirements to draw-down Registered Retirement Savings Plans (RRSPS) after age sixty-nine, clawbacks on the tax credit for persons age sixty-five and older, and income tax regulations that prohibit a person from accruing benefits in a defined benefit pension plan and drawing from that plan at the same time.

The arguments presented here suggest that those jurisdictions that have removed the age cap in their legislation[38] (and thereby provide protection against age discrimination at *all* ages) but that exempt bona fide retirement or pension plans, including mandatory retirement provisions (thereby allowing mandatory retirement in most circumstances) have it correct. This policy response ensures that persons of all ages have protection against age discrimination, but that private parties are allowed to agree to mandatory retirement, but only if it is accompanied by the security of a bona fide retirement or pension plan, or possibly a collective agreement. Allowing mandatory retirement if it is part of a collective agreement in the rare case that there is not a pension or retirement plan could be debatable, although there are protections to ensure that the concerns of all members are represented in the trade-offs involved. The general argument in favour of allowing mandatory retirement is bolstered by the fact that mandatory retirement tends to be associated with good jobs where it is reasonable to assume that the parties are well-informed and have the protection of a formal personnel practice or collective agreement.

In summary, four policy recommendations follow from this analysis:

1. Remove the age cap in the human rights codes that still have it so as to provide protection against age discrimination to persons of all ages.

2. Exempt bona fide pension and retirement plans, and possibly collective agreements that have a mandatory retirement age. This allows private parties to agree to mandatory retirement, but only if it is accompanied by the protection of a pension and retirement plan, and possibly a collective agreement.

3. Remove, or at least reconsider, the institutional and legal barriers that discourage, often unintentionally, the continued labour force participation of older persons and that discourage flexible transitions into and out of retirement.

4. Provide more information to all parties on issues that are likely to affect retirement decisions and the decision to agree to mandatory retirement, including information on the barriers that inhibit flexible retirement, experiences with later mandatory retirement ages, with voluntarily abandoning mandatory retirement, or more flexible mandatory retirement policies that permit re-contracting.

9

Challenging the Economic Assumptions of Mandatory Retirement

Jonathan R. Kesselman

Introduction

Nobel-prize-winning economist Wassily Leontief gave his 1970 presidential address to the American Economic Association the title, "Theoretical Assumptions and Nonobserved Facts." He expressed concerns about how complex mathematical economic models are often used in addressing issues of economic policy:[1]

> By the time it comes to interpretation of the substantive *conclusions*, the assumptions on which the model has been based are easily forgotten. But it is precisely the empirical validity of these *assumptions* on which the usefulness of the entire exercise depends.

Leontief's comments are devastatingly germane to how economic theory has been applied to the policy of contractual mandatory retirement over the past twenty-five years. Because it results from collective agreements and associated pension plans, mandatory retirement is "contractual" and thus supposedly voluntary, even though it forces some older workers to leave their jobs sooner than they desire.

Economically based arguments remain at the core of support for mandatory retirement policies, which continue to be permitted for most of the Canadian labour force outside Quebec and Manitoba.[2] Yet the key assumptions upon which the economic model of mandatory retirement hinges—some explicit and some implicit—have received little critical scrutiny in the public policy debate.

My analysis finds that the model traditionally used to support mandatory retirement has scant basis in terms of the realism or relevance of its key assumptions; it is based on postulated but "nonobserved" facts. Current policies permitting forced retirement, instead of promoting efficient labour markets, are economically costly to workers, employers, and public finances. Provinces other than Manitoba and Quebec permit mandatory retirement either by limiting coverage of their human rights codes to workers under age sixty-five or by explicitly exempting the practice in their codes.[3] Additional legislative or judicial bans on mandatory retirement would serve economic ends and eliminate a fundamental violation of older workers' human rights: judging individuals on the basis of their age rather than on their actual work performance. These findings are relevant to many Canadian provinces.

I begin by restating the claims made by proponents of the economic theory of mandatory retirement in scholarly research, judicial decisions, and employer usage. Then I investigate the behavioural assumptions that underlie the theory, focusing on the asserted voluntary nature of mandatory retirement and the perfect foresight of individual workers. Next, I critically assess the putative beneficial effects of compulsory retirement, including employment effects, worker productivity and training, and deferred compensation and incentives. Key implicit assumptions of the theory in these areas are that workers will continue working indefinitely if given deferred compensation and that there is no productivity-related self-selection by workers about when to retire. I consider the divergence between private and public perspectives on behaviour, calling into question the net fiscal and economic benefits of permitting contractual mandatory retirement. The preceding issues are instructively applied to an area often viewed as posing special problems for a ban on mandatory retirement—university professors and the tenure system. I conclude by examining some common characterizations of mandatory retirement and offering a summary of my findings.

The Claims of Economic Theory

Economic theory has been developed to explain mandatory retirement and to show the benefits that might arise for employers and workers from such agreements. American economist Edward Lazear produced the first formal model in 1979, the same year that University of Toronto economist James Pesando offered parallel intuitive insights.[4] In essence, the theory states that an employer offers a pattern of wages and pension benefits that underpays its workers during early years, relative to their productivity, and overpays them during later years. This system of "deferred compensation" gives employees incentives to stay with the firm for a long time, accept training, and apply themselves diligently. It also allows the employer to invest in job-specific skills with some assurance that workers will stay long enough for the firm to reap the returns, and it reduces the employer's costs of worker turnover, hiring, and monitoring.[5] By capping the period of "overpayments," it is argued, mandatory retirement allows for more efficient agreements to be reached. It is further asserted that mandatory retirement gives employers a non-disputatious means to terminate workers with declining productivity at age sixty-five. In Pesando's words, "banning [mandatory] retirement would lead to both the creation of costly new productivity measurements and a new arrangement of work rules, as well as the elimination of an effective dismissal procedure."[6]

Pesando and his fellow U of T economist Morley Gunderson were early Canadian proponents of contractual mandatory retirement on economic grounds. In 1988 they summarized the supporting arguments as follows:[7]

> Mandatory retirement is often justified on the grounds of opening up job and promotion opportunities for younger workers.... Mandatory retirement may facilitate planning on the part of both employers and employees. For employers, having a known

termination date to a particular contractual arrangement facilitates planning for replacements, pension payouts, training and upgrading.... For employees, a fixed retirement date pressures them to plan for their retirement and this, in turn, is likely to leave them better prepared for the eventuality of retirement ... A fixed retirement date minimizes the need to monitor and evaluate the performance of older workers.... [I]f mandatory retirement is banned, employers will have to monitor and evaluate the performance of their older workers more carefully. [E]mployers will be faced with the inevitability of having to dismiss some older workers.... [S]ome older workers may be dismissed prior to what would have been their mandatory retirement age, and others will be subject to more scrutiny and evaluation. This in turn may jeopardize their human rights and affect the notion of their retiring with dignity.... Mandatory retirement facilitates a deferred compensation system by providing a finite termination date to the contractual arrangement ... [S]uch a deferred compensation system ensures honesty and work effort on the part of employees because they want to be retained in order to receive the deferred compensation.

In 1990, the Supreme Court of Canada rendered judgment in a group of cases challenging the mandatory retirement practices of several universities. In the leading case of *McKinney v. University of Guelph*, the majority judgment echoed the economic argumentation of Gunderson and Pesando as follows:[8]

[Mandatory retirement] permits employers to plan their financial obligations, particularly in the area of pension plans and other benefits. It also permits a deferred compensation system whereby employees are paid less in earlier years than their productivity and more in later years, rather than have a wage system founded on current productivity. In addition it facilitates the recruitment and training of new staff. It avoids the stress of continuous reviews resulting from ability declining with age, and the need for dismissal for cause. It permits a seniority system and

> the willingness to tolerate its continuance having the knowledge that the work relationship will be coming to an end at a finite date. Employees can plan for their retirement well in advance and retire with dignity. Another important objective [is] the opening up of the labour market for younger unemployed workers. The problem of unemployment would be aggravated if employers were unable to retire their long-term workers. To put it in its simplest terms, mandatory retirement has become part of the very fabric of the organization of the labour market in this country ... The repercussions of abolishing mandatory retirement would be felt in all dimensions of the personnel function with which it is closely entwined: hiring, training, dismissals, monitoring and evaluation, and compensation.

This economic perspective continues to be cited by employers, as in a 2003 draft policy on post-retirement contracts by the University of British Columbia:[9]

> There are sound reasons for [UBC's] mandatory retirement policy. These include:
>
> - ensuring systemic opportunities for faculty renewal through the hiring and promotion of newly qualified faculty members....
> - creating an environment in which it is acceptable to have a deferred compensation system that involves lower pay in earlier years and higher pay in later years.
> - given the need for the University's tenure system, enabling individuals to retire from the University with dignity. Without mandatory retirement individuals who do not retire voluntarily must ultimately be dismissed for cause. This is undesirable and would necessitate detailed performance appraisals to justify each dismissal decision.

- encouraging and enabling faculty to plan in advance for retirement…

- enabling the University to plan for change and renewal.

Because of the influence and durability of the economic arguments used to support mandatory retirement, they warrant careful critical scrutiny. Without the defence of sound economic reasons, the clear violation of older workers' human rights by mandatory retirement would quickly lead to a ban on the practice. I next critically review the assumptions that underlie the economic arguments.

Behavioural Assumptions of Mandatory Retirement

A key assumption of the economic analysis is that all agreements for mandatory retirement at age sixty-five are fully consensual among the parties concerned.[10] As stated by Gunderson, mandatory retirement is "part of a private contracting arrangement between consenting parties … mutually agreed to by employers and employees who generally have considerable individual or collective bargaining power."[11] That "consensual" characterization might be accurate if all agreements were between an individual employee and the employer, such as the employment contract of an executive that specifies a termination date or age. However, almost all workers subject to mandatory retirement are covered by collective agreements rather than individual contracts. The linkage between mandatory retirement and pension plans, most often found in unionized firms, is also very strong. In a sample of older workers from the United States, 62 percent of workers with a pension plan had contractual mandatory retirement, and 86 percent of those with mandatory retirement had a pension plan.[12] In Canada, mandatory retirement is highly concentrated in work covered by collective agreements or formal personnel policies, not individual contracts.[13]

Once an organization like a trade union intermediates between the wishes of employees and the employer, the consensual nature of the relationship for the individual worker is broken.[14] Union officials may have a preference for mandatory retirement, perhaps as a way of opening positions for younger workers or as a way of cultivating support from members with many years ahead of them. Even if the union's leadership is fully responsive to its members, it will follow the preferences of its "median" voters. With respect to any issue, including the imposition of mandatory retirement, the union cannot satisfy the tastes of all its members. Those who would prefer to work beyond age sixty-five, typically a minority, are compelled to accept the terms negotiated for all members of the union at that establishment.

One might argue that workers who anticipate that they will want to continue working beyond age sixty-five should simply find jobs in non-union firms or firms that do not practice mandatory retirement. Yet they may work in an industry or occupation that is universally unionized, or they may work in a locale where there is one dominant unionized employer. Should such workers be expected to change their industry, occupation, or residence in order to find employment that does not bind them to retire at age sixty-five? That kind of mobility can impose large costs on individuals in terms of retraining or relocation or loss of work by a spouse. In addition, taking work with a non-union employer imposes a loss of wages and fringe benefits in most occupations.

Moreover, some groups may have a stronger preference for working beyond sixty-five than the average worker. Among American workers covered by mandatory retirement agreements, it was found that the specified retirement age was higher in agreements involving women than men and those involving blacks than whites.[15] Women who entered the labour force relatively late, after raising children or marriage breakdown, and recent immigrants may wish to work until later ages than "typical" non-immigrant male workers. These groups often need to work until a later age to accrue adequate pension benefits or savings for their retirement. These concerns are of particular salience to women, given their longer life expectancies. To the extent that these workers are employed alongside "typical" male workers who favour compulsory retirement, their ability to choose is constrained. It is telling that the two women justices took the minority side in the Supreme Court's 1990 decisions on mandatory retirement. They alone supported the elimination of mandatory retirement, and Madam Justice Wilson observed (at pg. 415–16),[16]

> [W]omen workers generally are unable to amass adequate pension earnings during their working years because of the high incidence of interrupted work histories due to child-bearing and child-rearing. Thus, the imposition of mandatory retirement raises not only issues of age discrimination but also may implicate other s. 15 rights as well [i.e., sex discrimination].

Even if all mandatory retirement were explicitly based on consensual agreements between individuals and their employers, one might question the ability of most people to predict their situation and needs many years into the future. The assumptions of perfect foresight and rational choice by younger workers—both essential for the optimality of contractual mandatory retirement—have been challenged.[17] Moreover, economic analysis now recognizes the possibility of myopic behaviour by younger workers in their lifetime savings and retirement planning decisions.[18] Most younger workers find it difficult to forecast their financial or health status or even marital and dependency status at age sixty-five. Many who might, when young, like to retire at sixty-five will find when they reach this age that their circumstances are quite different than anticipated. For example, some will find themselves far short of their lifetime savings goals on account of marital dissolution, child support payments, the costs of second families, unexpectedly low earnings, or simply poor investment returns. Others will find they continue to like work and its social connections more than anticipated.

Those individuals subject to mandatory retirement, yet who would like to continue working beyond age sixty-five, will be "involuntarily retired" from their current jobs. Typically, even in provinces which outlaw age discrimination beyond sixty-five, seniors face significant difficulties in finding work that fully utilizes their skills and experience. The time needed to find work after being laid off is usually much longer for workers aged 55 to 64 than for workers aged 25 to 54, and this disadvantage extends to workers terminated by mandatory retirement at age sixty-five. Older workers are twice as likely not to have a new job one year after layoff—63 percent versus 30 percent.[19] Older job seekers are also shut out of most positions that require much job-specific training by employers, who cannot count on a long enough period to recoup their investment in older workers. Most of these workers suffer a significant earnings decline in their new jobs, reflecting, job-specific skills and experience useful only to their previous employer.

In short, the behavioural assumptions needed to justify mandatory retirement are at best relevant to one-to-one bargaining between individual workers and their employers. The intermediary role of unions in this process disrupts the consensual nature of the agreements for many workers. Even when they can find other work that does not carry mandatory retirement, these individuals are likely to bear heavy costs. Younger workers who make consensual agreements to a fixed retirement age will often find their circumstances in later life changed in ways that they could not have predicted. Most workers who are forced to retire by mandatory retirement but wish to continue working will suffer large earnings losses in their alternative employment.

Asserted Beneficial Effects of Mandatory Retirement

Proponents of contractual mandatory retirement claim the practice yields significant positive benefits to workers, their employers, and the economy more broadly. A statement of these putative benefits in the contemporary Canadian context asserts that mandatory retirement serves to[20]

> ... open job and promotion opportunities for younger workers... [and] facilitate retiring with dignity and reduce the need for more constant monitoring and evaluation and possibly ultimate dismissal of older employees [and] enable deferred compensation [which in turn serves] ... purposes, such as reducing unwanted turnover and shirking, enabling retrospective and periodic monitoring, and encouraging worker commitment, loyalty and bonding to the company.

These benefits were promulgated in the original analysis supporting mandatory retirement, with the exception of the first item. Lazear intended his theory to explain the practice of mandatory retirement as self-interested, mutually beneficial behaviour by employers and their *current* employees; he rejected the "young worker" view.[21]

It is useful to assess all of these claimed benefits for allowing mandatory retirement. If benefits are found to be limited, absent, or even negative, then banning mandatory retirement could prove innocuous and perhaps even beneficial. I assess the claims under three headings: (1) employment effects, (2) worker productivity and monitoring, and (3) deferred compensation and incentives. My analysis finds the claims to be mostly postulated rather than proven, and the little evidence that exists to be sketchy, often inconclusive, and more suggestive of social costs than benefits for mandatory retirement.

Employment Effects

One of the most commonly cited arguments for mandatory retirement is that it is needed to open employment and promotion opportunities for younger workers. This point has been asserted by leaders in business, unions, and public sector institutions such as universities. For example, the president of the Ontario Federation of Labour was quoted as stating that mandatory retirement "provides job opportunities for young people coming into the workplace."[22] This "young worker" argument was part of the evidence accepted by the Supreme Court of Canada in its decisions upholding mandatory retirement. The majority judgment stated, "The problem of unemployment would be aggravated if employees were unable to retire their long-term workers ... [T]here is a significant correlation between those who retire and those who may be hired" (*McKinney* at pg. 295, 287). Nevertheless, the Court's decision did not give much weight to this point even though it was ruling in the context of universities where limited public funding would mean that deferred retirements could slow the hiring of young academics.

The young workers argument runs counter to elementary economic principles by assuming the economy offers only a given total amount of work—what economists call "the lump-of-labour fallacy." Job displacement may arise in the very short run, in narrowly defined occupations, or in recessionary periods, but over the long run the economy can create as many jobs as there are workers able and willing to fill them. Indeed, an economy's long-run growth is constrained by the availability of workers with the requisite skills, motivation, and experience. Moreover, young workers are hardly a substitute for the skilled and experienced workers who are forced to leave on account of mandatory retirement provisions. An early study of the effects of banning mandatory retirement in Ontario reached a similar conclusion: "The argument that ending compulsory retirement would reduce the job opportunities available in the labour force is not substantiated by economic analysis…."[23]

The "young worker" argument is particularly archaic under contemporary labour market conditions, and those forecast for future years. The Canadian economy is entering an era of skill shortages, even of workers with limited skills. Training of young workers and upgrading the skills of those already at work must be complemented by policies to encourage older workers to continue contributing their skills as long as possible. Mandatory retirement policies not only cut short the working lives of some older workers, they also cause some of the most productive middle-aged workers to depart to other jurisdictions that do not allow the practice. The Ontario Human Rights Commission cited this phenomenon arising at Ontario universities.[24]

Those who fear a flood of elderly workers were mandatory retirement abolished should consider the current situation. Two-thirds of workers choose to retire before age sixty-five, 43 percent retire before age 60, and the average age of retirement for all workers was 61 years in 1999, down more than two years since 1989. Still, 11.8 percent of the population aged 65 to 69 were active in the labour force in 2001, and more than one-fifth of all workers aged 45 years and older planned to retire after sixty-five or never to retire.[25] About half of Canadian workers are in jobs with mandatory retirement, and an estimated 6 to 20 percent of them would like to work beyond that age. Hence, an additional 3 to 10 percent of workers would choose to continue in their jobs were mandatory retirement banned.[26] However, the total employment effect of a ban on mandatory retirement is likely to be smaller, as some workers now forced to retire find other jobs or continue working on special terms with their original employer.

The policy-relevant questions are how many would continue to work, and how long they would work, if not constrained by the practice. The only recent empirical study using Canadian data found these effects difficult to estimate reliably, but suggested the total employment impact would be quite small. An earlier empirical study, undertaken several years after Manitoba and Quebec banned mandatory retirement, found the effects on labour force participation rates of people aged sixty-five and over to be statistically insignificant.[27] An American study of university professors found much larger impacts, not surprising in view of that occupation's high pay and job satisfaction and limited physical demands.[28] In comparative terms, the proportion of persons aged 65 to 69 who work is twice as high in the United States (which bans contractual mandatory retirement) as in Canada, though other factors may explain much of the difference.

Overall, the available evidence suggests abolition of mandatory retirement is unlikely to have a major impact on average retirement ages or years of work in Canada. Hence, any potential adverse effects on younger workers are minimal, or non-existent, but gains in terms of an enlarged skilled workforce and reduced public finance strains are similarly limited. To achieve the full benefits of an expanded workforce of older workers, abolition of mandatory retirement would need to be complemented by adaptations of other public policies and workplace practices.[29]

Related claims that mandatory retirement opens promotion opportunities for younger workers—and that banning the practice would significantly impede promotions—are equally flawed. Even if this did not run afoul of the lump-of-labour fallacy, the size of the effect would necessarily be small. Evidence presented below suggests a ban on mandatory retirement would raise the average retirement age for the overall labour force by just a fraction of one year—most likely under four months. Hence, such a ban would at most simply defer the average age for promotions of younger workers by several months. That hardly constitutes a significant blockage for the promotion of younger workers, although the delays could be longer in some occupations. The delays could also be lengthened in the transition period from a situation with mandatory retirement to a new "steady state" without it.

Worker Productivity and Monitoring

A central argument for mandatory retirement is that the productivity of workers declines as they approach age sixty-five and in the years beyond.[30] Advocates argue contractual mandatory retirement allows for graceful discharge of workers with no need for harsh and costly performance monitoring to determine when they are no longer productive.[31] Indeed, in the *McKinney* case, the Supreme Court justices accepted evidence that "on average there is a decline in intellectual ability from the age of sixty onwards," (at pg. 289), which they used to justify accepting age sixty-five as a basis for allowing mandatory retirement. It is ironic that the mean age of the justices deciding the case was sixty-five, three were over sixty-five, and Supreme Court justices can continue holding office to age seventy-five. This irony is heightened by the fact that the case concerned the forced retirement of professors, whose work entails physical and intellectual demands not unlike those of Supreme Court justices.

Empirical studies of the determinants of individual worker performance provide little evidence that ability or productivity declines with age, except in occupations that require physical strength or sensory acuity.[32] The typical pattern is for a worker's productivity to rise over most of the lifetime, then level off for the last ten or fifteen years. Even if an individual worker's skills do eventually decline with advanced age, there is no evidence that this occurs abruptly at sixty-five, or as early as sixty-five, in most occupations. In white-collar work, one study reported that "reliability and especially experience seemed to compensate for the effects of somewhat reduced physical abilities."[33] Workers who experience declining physical, sensory, or mental faculties are more likely to retire voluntarily prior to age sixty-five. Current high rates of early retirement reflect this self-selection behaviour, as well as responses to the financial incentives in private and public pensions and taxes.

Ordinary economic behaviour ensures that workers with declining abilities will be the ones most likely to choose early retirement. Individuals' labour force participation and retirement decisions can be modelled as a utility-maximizing choice.[34] Utility is derived from income from all sources (including earnings, savings, and pensions), from leisure time, and from job satisfaction. It is natural that the utility of working will decline for individuals as their health status declines, as work becomes more demanding, and as their productivity and job satisfaction fall. Some will face declining earnings, but even at unchanged pay levels, they will be more inclined to choose early retirement. As noted in a Labour Canada report, "poor performers are usually unhappy in their jobs and are anxious to leave as soon as feasible."[35]

Advocates of mandatory retirement argue further that banning the practice would necessitate costly systems to monitor the performance of older workers, catching those who wish to stay on too long. The self-selection process will limit the numbers of older workers who fall into this category. Moreover, given their age, the work period ahead of them is limited in any event. Employers need to be more concerned about *younger* workers whose productivity is poor or declining. Those workers have far more years ahead of them in which they might constitute a burden to their firms, and they do not have the financial options of public and private pensions to leave the workforce early. In short, employers need effective processes to monitor the productivity of workers at all ages, and little if anything additional would be needed for older workers in the absence of forced retirement. Despite the claims, there is no evidence of costly new monitoring processes in Canadian jurisdictions that have banned the practice.

Deferred Compensation and Incentives

Another key aspect of the theory and claims relating to mandatory retirement is that there must be a terminal point for the worker's employment in order for deferred compensation to provide effective incentives. Deferred compensation is the practice of paying workers less than their productivity in earlier years and more than their productivity in later years with the firm. This deferral can be achieved by a steeply inclined age-earnings profile and by pensions and post-retirement benefits that grow with the worker's tenure. The professed purpose of deferred compensation is to augment workers' incentives to work hard so as not to lose the deferred compensation if they are "shirking" and get fired, and incentives to remain with the same firm for a long time, so employers can reap returns from investments in training.

Empirical research offers mixed findings on the reality of deferred compensation, though it does appear to be present in skilled occupations. While some studies are supportive of deferred compensation, others find that both productivity and compensation rise in lockstep with the worker's age.[36] An age limit on employment is not crucial to this process. Proponents of mandatory retirement have asserted, "Without such a termination date, compensation could exceed productivity *indefinitely* and hence the contractual arrangement *could not exist*."[37] This claim is puzzling, since workers continuing beyond age sixty-five would not work "indefinitely;" for most it would be no more than a few additional years. The previously cited fact that the majority of workers choose early retirement casts further doubts about the salience of a cutoff age.

For example, if all workers now constrained by mandatory retirement who wish to continue beyond age sixty-five (about 3–10 percent of the labour force) were to work an additional three years, this would add only about one to four months to average working lives.[38] That seems hardly enough to upset deferred compensation arrangements and the putative associated efficiency gains. If deferred compensation were one-third of the worker productivity (or one-fourth of total compensation) in later years, adding one to four months to the average working life would constitute just 2 to 8 percent of one year's earnings. Averaged over the last 20 years of work in which there is positive deferred compensation, the impact is a trifling 0.1 to 0.4 of 1 percent of total compensation for that period. Also, based on my earlier analysis, those choosing to work longer would typically be workers with above-average productivity, so the extent of any deferred compensation would be minimized.

If in fact there were any adverse effects, deferred compensation could adjust to the abolition of mandatory retirement in many ways. The age-earnings profile could be flattened or even decline beyond a specified period for workers with below-average productivity, based on performance reviews. Various adjustments to public and private pension schemes could be implemented, and early retirement incentive schemes modified. Mandatory retirement was abolished in the United States in 1986, but there has been no ensuing evidence of reduced incentives to work hard or to stay with an employer. If deferred compensation schemes did play a role in these incentives, suitable adjustments to those schemes must have occurred. Indeed, empirical analysis suggests that age discrimination laws, akin to but broader than a mandatory retirement ban, have actually strengthened long-term incentive contracts by steepening age-earnings profiles in the United States.[39]

One might also question whether deferred compensation is as useful an incentive device for the purposes of firms training their workers as in earlier years. Given the more rapid obsolescence of workers' skills, training has become more a recurring process than a one-time event at the start of a worker's tenure. Hence, returns to a firm's investments in training must be reaped more quickly, and the role of deferred compensation to purchase lifetime loyalty will have diminished. This point also means a firm can invest in training its older workers with some expectation that it will obtain an adequate return before they retire. Additionally, the declining lifetime security of jobs in many occupations weakens the role of deferred compensation in labour market incentives. The economic theory of contractual mandatory retirement assumes that workers spend all or almost all of their working lives with a single employer; in fact, very few workers retire from the same employer that hired them when they first began full-time work.

Public Versus Private Perspective

Even if the mandatory retirement process were not deficient on the grounds already cited, it has further problems from a public policy perspective. The practice of mandatory retirement could be socially optimal if the parties directly involved—employers and their employees—captured all the benefits and carried all the costs of their agreements. Mandatory retirement also imposes costs on the rest of society, so these private, voluntary agreements may not be in the public interest. External costs take the form of fiscal impacts (on tax revenues and public expenditures) and economic impacts arising through the tax burden and the labour market. Because of these external costs not recognized by the parties to mandatory retirement agreements, society may have an interest in banning or at least restricting the practice.

Fiscal Impacts

With mandatory retirement, some workers who would like to continue working beyond age sixty-five are involuntarily terminated from their jobs. While some workers may find alternative work after being forced to retire, that work often pays significantly less, and others will not succeed in finding work at all. Hence, many persons terminated on account of mandatory retirement will pay less in income, sales, and payroll taxes, and begin drawing public pension benefits sooner than if they had been allowed to continue working. The adverse impact on tax revenues would be offset if the forced retirements opened up equivalent, equal-paying jobs for unemployed workers; my earlier analysis suggests that this is not the case. Taken together, the decreased tax revenues and increased public pension costs from mandatory retirement will aggravate the fiscal strains that will arise in future years from the bulge of retiring baby boomers.

On the tax side, part of the reduced earnings from those involuntarily terminated will be offset by earlier withdrawals of taxable funds from tax-deferred savings plans such as company pensions and RRSPs. Yet those additions to the tax base represent simply an acceleration of funds that would have been withdrawn and taxed in later years, had those workers been allowed to continue. Since taxable withdrawals will also be less than the lost earnings, mandatory retirement exerts a negative net impact on the taxable base. Were mandatory retirement banned, it would be desirable to raise the age for mandatory distributions from tax-deferred savings plans above the current sixty-nine; alternatively, individuals' mandatory distributions could be reduced by the amount of their labour earnings.

Contractual mandatory retirement can also increase fiscal strains on the expenditure side of federal budgets.[40] For applicants aged sixty-five and over, the Canada Pension Plan (CPP) does not impose any work or earnings test in order to receive benefits. For delayed start of benefits beyond the standard age of sixty-five, the benefit payment is increased by 0.5 percent per month up to the age of seventy, but this provision under-compensates deferred beneficiaries relative to the actuarial cost. Hence, workers who begin their CPP benefits earlier than they would have preferred on account of mandatory retirement cost the program money. The Old Age Security (OAS) pension is paid at age sixty-five irrespective of work, but the reduced earnings for some workers caused by mandatory retirement means there is less tax clawback of OAS payments from higher earners. The largest impact of mandatory retirement on budgetary costs arises through the income-tested Guaranteed Income Supplement (and the companion Allowance) program. Some who would have continued working beyond sixty-five will be forced to retire and find their incomes decline to a point that they draw GIS benefits.

Additional costs are imposed on provincial budgets through higher health care costs for workers compelled to retire.[41] A variety of medical and dental services and prescription drugs are provided on an income-tested basis by provincial programs (often on more generous terms for seniors). Workers forced to retire are more likely to have incomes fall to the levels at which such benefits are covered. There is also growing evidence that physical and mental inactivity can spur the degenerative diseases that afflict many older people. One study that overcame the problem of isolating causation found that increased labour force activity by seniors reduced mortality rates. Another study found that persons who were forced to retire earlier than they wished experienced a lower quality of life than those who had a choice in their retirement.[42]

One might question whether the fiscal impacts of people's work and retirement decisions are relevant to public policy. Shouldn't policy respect people's preferences and eschew attempts to prolong working lives? From the public policy perspective, the appropriate reference point is the work and retirement choices individuals would make in the absence of distortions imposed by taxes, public pensions, and regulation of private pensions. However, given that most such public policies are tilted toward incentives for shorter work lives and early retirement and against longer work lives and phased retirement, the fiscal impacts are relevant to sound public policy. The fiscal impacts of biases from mandatory retirement practices that shorten working lives relative to individual preferences are similarly germane for policy formulation.

Economic Impacts

To the extent that mandatory retirement lowers tax revenues and raises public expenditures, this practice puts upward pressure on tax rates. Higher tax rates increase the distortions of the economy and can reduce its efficiency and growth, though the extent of this damage depends on the exact form of the taxes that are increased. The upward pressure on tax rates from mandatory retirement will only compound other powerful stresses on public finances and tax rates over the next generation. The demographics of the baby boom and bust will radically alter the economic landscape. For the first time in Canadian history, retirees from the labour force will outnumber new entrants. From the current four workers per senior, the balance will shift to just two workers per senior by mid-century.[43] Allowing a practice such as mandatory retirement, which exacerbates these problems, runs counter to desirable public policy.

Contractual mandatory retirement can have a more directly adverse impact on economic performance by withdrawing some of the most skilled and experienced workers. Even when those individuals can find alternative work with another firm, that work typically does not exploit the full firm-specific skills and knowledge that they had acquired over years in their previous job. Older workers bring to the workplace a wealth of institutional knowledge and networking connections that cannot be replaced by younger workers. While their current employer could rehire these workers under a special contract immediately after retiring them, often tradition or institutional rigidities (including collective agreements) stand in the way. As the Canadian economy enters an era of growing skills shortages, any policy that allows the most seasoned workers to be forced out of their jobs seems counterproductive.

Just as women were the major driver in the Canadian labour force over the past two generations, older workers have the potential to play a similar role in the future. It is forecast that by 2010 fully 70 percent of the net increase in the working age population from 2000 will arise in ages fifty-five to sixty-four, and by 2020 all of the net increase will be in that age group.[44] Moreover, growth in the population aged sixty-five and above is perched to accelerate rapidly. It makes little sense that average retirement ages have declined at the same time that life spans have risen, health status of older persons improved, and the physical demands of most jobs fallen. Increased time spent in formal education before starting work has further truncated the period of productive work. A person entering the workforce at age 22 and retiring at 61 spends just 39 years at work, barely half a lifetime. As argued by the Organisation for Economic Co-operation and Development (OECD), a prime challenge for policy will be to raise the average age of retirement to better balance people's economically active lives and their period of retirement.[45]

Demographic trends, in addition to improving health and life expectancies, are rapidly becoming more favourable for the extended employment of older workers. Canadians have been increasing their average educational levels, most notably for younger cohorts but progressing over time to older cohorts. In 1990, almost two-thirds of the population aged 55 to 64 had less than high school completion, but by 2010 this figure will fall to just one-quarter. In 1990, only one-fifth of those aged 55 to 64 had a college, university, or technical degree, a figure that will jump to nearly one-half in 2010.[46] Similar advances in educational attainment will apply to the group aged sixty-five to 74 ten years later, in 2020. Hence, failure to undertake policies that reverse early retirements and that allow individuals to continue working beyond age sixty-five will have much more adverse economic impacts than in the past.

Mandatory Retirement and Post-Secondary Faculty

Post-secondary faculty members offer a demanding test of my critique of the economic reasoning used to support mandatory retirement policies.[47] Professors are the ultimate "knowledge workers" and personify a sector of the labour force growing in relative size and economic importance. Their work entails long training and experience, high self-direction, little direct supervision, long-run assessment, and high pay and job satisfaction. When mandatory retirement was banned in the United States, professors extended their average working lives more than workers in other occupations.[48] Yet professors enjoy another condition of employment that makes it harder for their employers to terminate them if their performance is found to be flagging—the tenure system. This point was cited by the Supreme Court's majority decision in the *McKinney* case:

> Mandatory retirement is ... intimately tied to the tenure system which undergirds the specific and necessary ambience of university life and ensures continuing faculty renewal, a necessary process in enabling universities to be centres of excellence on the cutting edge of new discoveries and ideas (at pg. 234).

Legislative actions banning mandatory retirement have reflected concern for the situation of universities and their faculties. The United States passed the Age Discrimination in Employment Act in 1986 but, in response to pressures from college and university administrators, allowed the continued mandatory retirement of faculty members at age seventy.[49] Congress terminated this exemption in 1994, prohibiting mandatory retirement for faculty at U.S. post-secondary institutions. When Manitoba passed legislation to ban mandatory retirement in 1982, university faculty members were included. However, the 1996 amendments to the University of Manitoba Act exempted universities from the ban on the basis that age is a "*bona fide* occupational requirement" (BFOR) for academics.[50] Subsequently, two of the three Manitoba universities instituted mandatory retirement under collective agreements.

Typically, BFOR exemptions from mandatory retirement are restricted to occupations where individual abilities cannot be reliably assessed, and where errors of judgment or action can carry catastrophic or irreversible consequences. Examples include airline pilots, firefighters, police, and judges; exemptions for occupations of these kinds are found in federal and provincial legislation even where mandatory retirement is otherwise prohibited. It is doubtful whether the errors of omission or commission of an occasional bumbling professor who clings to office too long carry any such dire consequences. Ten years after Harvard University abolished mandatory retirement for faculty, students expressed their support for older professors:[51]

> Simply put, older professors are some of the most valuable teachers at Harvard. The faculty who choose to keep working late into

their lives do so out of passion. Often they love to teach and excel at it; others stay on because their eclectic research interests still awaken curiosity even after decades of study; and for many more it is a combination of the two. For undergraduates, the chance to interact with these dynamic professors and hear first-hand about their storied careers is one of the most exciting and unique aspects of being a Harvard student. We are grateful to these professors for dedicating their lives to academia.

My earlier challenges to the economic assumptions of mandatory retirement can be tested in the context of retirement for professors. First, the issue of making way for younger faculty members will have diminishing significance as growing shortages of candidates with the requisite training and abilities become universities' main problem. Indeed, the application of mandatory retirement in most Canadian universities has become a competitive disadvantage, as high-performing younger and middle-aged professors seek out jurisdictions where they will be allowed to continue working beyond age sixty-five. Second, those professors who would choose to work longer tend to be the most energetic and productive individuals, following my economic analysis of self-selection and as confirmed by Harvard University students' observations. This point receives further support from a study finding that professors whose work is weak or unsatisfactory are more likely than others to retire early even with mandatory retirement.[52] Third, a growing contingent of the professoriate is coming from women and recent immigrants, both tending to begin their academic work lives in Canada later in life. For women, this delayed start often results from the combination of years taken for child rearing plus the time for completion of graduate studies.[53] For both women and immigrants, the ability to work beyond sixty-five is often needed to accrue adequate pensions.

Deferred compensation, a key element in the economic theory of mandatory retirement, also plays a problematic role in the academic setting. The market for academics (like other professionals, business executives, and workers with advanced technical and creative skills) is national and international. If universities in provinces that allow mandatory retirement attempt to depress salaries for younger faculty and boost salaries for senior faculty, they will find it difficult to attract promising junior candidates. They are pressed to imitate the flatter lifetime salary profiles economic theory predicts will prevail at institutions not practicing contractual mandatory retirement.[54] Hence, there is little scope for some universities to pursue greater deferred compensation than the general pattern, which for top academics is set by American institutions without mandatory retirement. This point also challenges the implicit assumption that workers will spend all, or almost all, of their lives with a single employer. In the highly mobile market for academics, particularly the top talent, the more typical pattern is multiple employers over a lifetime.

In its majority judgment in *McKinney*, the Supreme Court suggested that a ban on mandatory retirement at universities would be incompatible with the tenure system. This view was based on the assumption that many older professors would need to be dismissed but tenure would prevent such dismissals. In a dissenting opinion Madam Justice L'Heureux-Dubé asserted (at pg. 244–45),

> There is no convincing evidence that mandatory retirement is the quid pro quo of the tenure system. The value of tenure is threatened by incompetence, not by the aging process. The presumption of academic incapacity at age sixty-five is not well founded. The discrepancies between physical and intellectual abilities amongst different age groups may be more than compensated for by increased experience, wisdom and skills acquired over time. There is therefore no pressing and substantial objective addressed by the mandatory retirement policy.

In fact, despite the abolition of mandatory retirement at universities more than two decades ago in Quebec and one decade ago in the United States, tenure systems have remained in place at almost all such institutions and continue to function well. This situation suggests no inherent conflict between the tenure system and a prospective ban on mandatory retirement for faculty members.

University administrations have diverse tools at their disposal for identifying and releasing faculty members, both over and under age sixty-five, who perform poorly. Individual salaries can be frozen, and thus reduced over time in real terms. Many institutions already weed out low performers through selective incentives for early retirement, and these can be applied to older faculty members as well as those under sixty-five. For faculty whose research has flagged but who continue to teach effectively, work can be shifted toward greater teaching, thus releasing younger faculty to undertake more research. Similarly, many older faculty members are highly effective as administrators and can assume more of that task from younger colleagues. Universities without mandatory retirement can also assist older professors in assessing their role and in making retirement plans; for example, Harvard University uses "discreet one-on-one advising" to counsel older professors and work out interim plans, often with phased retirement and continued access to office and library facilities.[55]

Thus, even in the restrictive setting of university professors and the tenure system, the economic arguments for retaining mandatory retirement do not hold up. Adequate provisions already exist for assessing performance, for shifting workloads where appropriate, and for inducing workers to retire where necessary. Contrary to the fears about disputatious firings cited by supporters of mandatory retirement, there are methods for instituting phased retirements, continuing university connections, and managing departures with dignity. In the relatively rare cases of malfeasance or incompetence where an individual resists moves to terminate, older faculty members do not raise any issues not already faced by university administrations for younger faculty. Evidence from Quebec and the United States is that professors over sixty-five do not present any more problems than their younger peers. Also, older professors depart more willingly in view of their larger accrued pension entitlements.

Characterizations of Mandatory Retirement

Those who support contractual mandatory retirement advance various characterizations to defend the practice. These ways of viewing mandatory retirement are outside the formal economic model but not entirely unrelated. One common critique is that a ban on mandatory retirement is just the thin edge for further policy changes to weaken both private and public pensions. If workers can hold their jobs indefinitely, it is argued, employers and governments will seize the excuse to reduce or delay pension benefits. This point is expressed most often by trade unions, which have fought hard for existing pension provisions.[56] Yet the issue of delaying the age for benefit entitlement, most often proposed for public pensions, is fully separable from the issue of compulsory retirement. Nevertheless, if extending productive working lives is seen as a desirable goal in the context of an aging population, healthier older workers, and less physically demanding jobs, one could argue that mandatory retirement bans and pension reforms can play useful complementary roles.[57] For employer pensions, a ban on mandatory retirement could even lead to enriched benefits or early retirement incentives as a way to induce the departure of less productive older workers.

Supporters of contractual mandatory retirement have offered some colourful analogies in seeking to rebut the arguments against the practice.[58] They contrast mandatory retirement with legally prohibited labour activities such as prostitution and indentured service, depicting mandatory retirement as relatively innocuous and analogous to marriage and mortgage contracts. More apt analogies would be the extensive regulations of labour markets in Canada and most modern economies. Individuals are banned from contracting with employers to work for less than a specified wage, for unduly long hours, under harassing conditions and in hazardous workplaces—even if the worker and employer could reach mutually agreeable terms. A ban on mandatory retirement is no different in kind from these commonplace labour restrictions, well-established for clear social or economic purposes.

Questions have been raised about the singling out of mandatory retirement among all the areas that can enter collective agreements based on a simple majority vote of the union members. Unionized workers and their employers typically agree to terms of employment that do not have the support of every last member. Yet with mandatory retirement, a collective agreement or pension plan is violating one of the Charter's enumerated protected characteristics: age. If an agreement specified that no women or visible or religious minorities were to be hired or promoted, this would equally violate the Charter protections.[59] These types of terms constitute vitally different matters than the wage scales, job classifications, work hours, and benefit provisions that are the heart of collective agreements. A ban on mandatory retirement would not imply any reduction in the scope of collective bargaining in the core areas or a need for unanimity in these decisions.

Supporters of mandatory retirement also argue that age differs fundamentally from the other enumerated characteristics; after all, with good fortune we all attain age sixty-five, but few change our sex or race. Even the Supreme Court expressed tacit acceptance of this point in its landmark ruling on forced retirement.[60] But this view ignores the fact that people age at different rates both physically and mentally; some workers are more vital and productive into their seventies and beyond than others are in their fifties. Protection against age discrimination in employment is based on the premise that individuals should be judged on their own performance, not on an extrinsic attribute such as age (or sex or race) or on stereotypes concerning older workers. Hence, age should be treated no differently than the other enumerated characteristics in the Charter. The Supreme Court ultimately agreed that mandatory retirement violates section 15 of the Charter: prohibition against discrimination on the basis of age.[61]

Summary of Findings

My analysis finds many of the key assumptions underpinning the economic model of contractual mandatory retirement to be unfounded or in the words of Leontief, based on "non-observed facts." Accordingly, the economic case for mandatory retirement fails on several grounds:

- The model of mandatory retirement characterizes the practice as voluntary, utility-enhancing agreements that must bring benefits to both employers and workers. Yet mandatory retirement provisions usually are not agreements between individual workers and their employers but rather the result of collective agreements. Hence, some individuals (especially women and recent immigrants) may wish to work longer than the median union voter, but they will be constrained by compulsory retirement. Additionally, in some provinces the employer can unilaterally impose the age of retirement, which is totally at odds with the model's assumptions.

- The model further assumes that individual workers have perfect foresight about their tastes and situation throughout their lives, and can thus commit to a fixed retirement age early in working life. In fact, many individuals will find later in their working life that they have not accurately predicted their family, health, income, or wealth situation, and will wish to alter their planned retirement date, often extending work beyond their original plans.

- Economic theory asserts that mandatory retirement facilitates deferred compensation schemes that bring significant economic benefits in worker loyalty and diligence. It further implies that a ban on mandatory retirement would be costly to employers in terms of increased monitoring, turnover, and training of workers. Yet jurisdictions that banned mandatory retirement many years ago—Manitoba, Quebec, and the United States—have observed little, if any, of these predicted adverse consequences.[62]

- Economic theory also asserts that with deferred compensation, individuals would choose to work "indefinitely" to capture more wages that exceed their actual productivity. In fact, even in employment offering mandatory retirement and deferred compensation, a substantial majority of workers choose to retire well before age sixty-five. The theory additionally ignores the self-selection process in which less productive workers tend to retire earlier and more productive workers tend to retire later.

- A recurrent economic argument for mandatory retirement is that the practice opens up work and promotion opportunities for younger workers. But this runs counter to what economists call the lump-of-labour fallacy; a healthy economy is actually limited in its growth by the availability of workers of all ages. Moreover, the looming shortage of skilled workers means that the economy would benefit from allowing workers to stay on the job longer.

- The standard economic case for mandatory retirement ignores the social costs that parties to these agreements do not consider but that are essential to a public policy perspective. These costs include the impacts of workers retiring earlier than they desire on tax revenues, public pension and health costs, and the economy. With the impending fiscal stresses from an aging population and a shrinking workforce, these impacts argue further for abolishing mandatory retirement.

- As a relatively stringent test of the mandatory retirement model assumptions, the case of universities and the tenure system was examined. Even in this context, the model's assumptions are found to be deficient, and academic employers are seen to have sufficient human resource tools to deal with a ban on forced retirement. Both the tenure system and employer pension plans have survived at Quebec and U.S. universities, and there is no indication of declining overall academic quality.

A ban on mandatory retirement would simply extend an existing long list of prohibited types of labour agreements. It would not undermine unions' ability to bargain over diverse terms of employment, nor would it constrain collective agreements to unanimity of union members. Forced retirement is similar to other forms of employment discrimination (such as that based on sex or race) in judging workers on their extrinsic characteristics rather than their individual abilities and work performance.

The case for allowing mandatory retirement to continue is based on economic analysis that presumes markets always produce desirable results. One proponent has argued:

> Since the current package of rules [including contractual mandatory retirement] has evolved from market forces and without legislated restraints, economic theory prescribes that [a ban on mandatory retirement] must entail a net loss in efficiency.[63]

The Supreme Court accepted this presumption in its landmark rulings to permit age discrimination on the basis of the asserted economic benefits. Yet while racial and sexual discrimination were once the norm in Canada, economic analysis has shown how discrimination can persist in competitive markets but still fail tests of efficiency.[64] Legislative bans have reduced these unwarranted forms of discrimination and also stimulated incentives for advanced education and labour force activity by groups that have made increasing contributions to the economy. The same point could be made about older workers, age discrimination, and a prospective ban on forced retirement.

Mandatory retirement, even in contractual agreements between workers and employers, cannot be justified on economic grounds. The economic theory underpinning mandatory retirement hinges on a series of unrealistic assumptions, thus undermining its policy prescriptions. As witnessed in jurisdictions where mandatory retirement has been banned for many years, this is not a radical change, nor have the alleged adverse effects emerged. Hence, such agreements should not be exempted from the prohibition of age discrimination on the basis that they constitute "reasonable limits prescribed by law as can be demonstrably justified in a free and democratic society."[65] Given emerging labour shortages, looming public finance stresses, and newer economic evidence, the Supreme Court decisions on mandatory retirement will inevitably be contested. Indeed, one lower court decision has already urged such a legal challenge.[66] But there is no reason for the provinces to await judicial action before proceeding. The remaining provinces should follow the lead taken more than twenty years ago by Manitoba and Quebec in banning mandatory retirement in both the public and private sectors.

10

The Debate about Mandatory Retirement in Ontario Universities: Positive and Personal Choices about Retirement at 65

John Munro

Introduction

On May 29, 2003, the Ontario Progressive Conservative government introduced a bill designed to eliminate contractual mandatory retirement.[1] The purpose of this bill was to prohibit private employers, and unions, from imposing mandatory retirement or from including such provisions even in freely negotiated labour contracts. However, the legislation was not passed before the government lost the election of October 2003 and was replaced by a Liberal administration. In August 2004, it declared its intention to introduce a similar bill, to be preceded by public hearings on this issue, in September 2004.[2] That the former Conservative government chose to introduce such a bill was a major surprise, for it had seemingly turned a deaf ear to pronouncements from the Ontario Human Rights

Commission to abolish mandatory retirement as a clear violation of basic human rights.[3]

The Ontario Human Rights Code, in the version enacted in 1981, explicitly states in section 5(1) on employment that:

> Every person has a right to equal treatment with respect to employment without discrimination because of race, ancestry, place of origin, colour, ethnic origin, citizenship, creed, sex, sexual orientation, age, record of offences, marital status, same-sex partnership status, family status, or disability.

But the following section 9(a) severely qualifies the ban on age discrimination by stating that in "subsection 5(1) ... 'age' means an age that is eighteen years or more and less than sixty-five." Furthermore, section 24(1)(b) contains yet another significant qualification:

> The right under section 5 to equal treatment with respect to employment is not infringed where ... the discrimination in employment is for reasons of age, sex, record of offences, marital status or same-sex partnership status, if the age, sex, record of offences, marital status or same-sex partnership status of the applicant is a reasonable and *bona fide* qualification.[4]

The crucial issue is whether or not provincial human rights codes should, in this fashion, *permit* age discrimination, specifically in the form of mandatory retirement for those sixty-five (and over) in Canadian universities and similar organizations, when section 15 of the Canadian Charter of Rights and Freedoms (1982) expressly forbids any such form of age discrimination.[5]

If mandatory retirement is indeed a form of unwarranted and unjustified age-discrimination, we must note that American legislation on this issue has long been more advanced than Canadian. From as early as 1967, the U.S. Congress has enacted a series of laws to protect the rights of older workers and finally to abolish mandatory retirement completely. In 1978, Congress amended the 1967 Age Discrimination in Employment Act to establish seventy (rather than sixty-five) as the minimum age of mandatory retirement for most workers; but university professors were then excluded (i.e., with mandatory retirement at sixty-five), until the act was amended in 1982. In October 1986, Congress prohibited mandatory retirement everywhere, though again with the significant exception of university professors—whose employment, however, was still permitted to continue until age seventy, i.e., five years beyond the standard age of mandatory retirement in Canada, at sixty-five. As Ashenfelter and Card inform us, that exemption "was a hard-fought victory for college and university representatives, who argued that mandatory retirement was needed to maintain a steady inflow of younger faculty and promote the hiring of women and minorities"—arguments that will be encountered later in this study.[6] The 1986 act had required Congress to review this exemption in seven years. It did so, and voted to have this exemption expire, on schedule, on December 31, 1993. By that date, many American universities and colleges had already, quite voluntarily, withdrawn their mandatory retirement provisions.[7]

For this issue in Canada, 1986 was also a significant year. Following the spirit of section 15 of the Charter, which had come into force the previous year (1985), the federal government abolished mandatory retirement for its own civil service employment. It was not, however, the first government in Canada to do so. Manitoba was the first, but by a rather circuitous route. In June 1974, the New Democratic Party (NDP) government of Edward Schreyer passed its own human rights act, one that also prohibited age discrimination, but—in contrast to all other provincial human rights legislation—without setting any age limits.[8] Subsequently, in 1980, Imogene McIntire, Professor of Education at the University of Manitoba, decided to contest the mandatory retirement that she faced that June, at age 65, by filing a suit with the Court of Queen's Bench to request a declaratory judgment that the university's retirement policy was in contravention of the 1974 Human Rights Act. The Court fully agreed (1981) that, because this Act lacked any upper limit on age, "no employer may refuse to continue to employ a person solely on the basis of his age."[9] In March 1981, the Conservative government of Sterling Lyon established the Commission on Compulsory Retirement under Marshall

Rothstein; and in that same year the Manitoba Court of Appeal upheld the *McIntire* decision, so that when Rothstein finally issued his report in 1982, he endorsed the abolition of mandatory retirement. As Thomas Flanagan comments in his study of this case, "[B]y 1982, the result was effectively the same as if the legislature had decided to abolish fixed-age retirement," even though, as he further contends, neither the NDP nor Conservative governments had ever favoured such an abolition.[10] The story does not end here. For, in 1996, following a successful, though bitter, faculty strike at the University of Manitoba, Gary Filmon's Conservative government amended its legislation governing the Universities of Manitoba, Winnipeg, and Brandon, to permit them to include contractual mandatory retirement in their contracts, with the argument that age is a bona fide occupational requirement for academics.[11]

Thus, contrary to often-expressed views, Quebec is currently the only province that has included university professors in its complete ban on mandatory retirement, a ban that was enacted in December 1983, in the bill known as "Respecting Labour Standards."[12] In 1982, New Brunswick had similarly enacted legislation ostensibly to abolish mandatory retirement, enabling those forced into retirement to file a complaint under the human rights code. Excluded from this provision are those whose contractual employment benefits contain a registered pension plan contingent upon retirement at a certain age (normally sixty-five). In neighbouring Prince Edward Island, mandatory retirement has been in force in universities since 1995. Finally, Alberta, despite having abolished mandatory retirement for its civil service in the early 1980s, and despite including a clause prohibiting age discrimination in its human rights code, still permits its universities to include mandatory retirement in contracts with their faculties (Universities of Alberta, Athabaska, and Lethbridge, but not yet Calgary). An important test case was resolved in September 1992 when the Supreme Court of Canada, in the decision known as *Dickason v. University of Alberta*, upheld that university's right to impose mandatory retirement, as stipulated in a contract with its faculty association.[13]

McKinney v. University of Guelph: The Role of the Canadian Supreme Court
Many of the arguments put forward in deciding the *Dickason* case are similar to those contained in the landmark legal decision, *McKinney v. University of Guelph*, issued by the Supreme Court of Canada in December 1990.[14] In 1985, Professor David McKinney of the University of Guelph, joined by eight other professors (and one librarian) at Laurentian University, York University, and the University of Toronto, supported by their faculty associations and the Ontario Confederation of University Faculty Associations (OCUFA), but opposed by their universities and the Attorney General of Ontario, filed a lawsuit to apply "for declarations that the universities' policies of the [sic] mandatory retirement at age sixty-five violate s. 15 of the Canadian Charter of Rights and Freedoms, and that s. 9a [now 10(1)] of the [Ontario] Human Rights Code, 1981, by not treating persons who attain the age of 65 equally with others, also violates s. 15." In 1989, the Court of Appeal for Ontario rejected their suit, which was then appealed to the Supreme Court of Canada.

In upholding the Ontario Court of Appeal's decision, the Supreme Court based its verdict on three essential grounds.[15] First, the Charter of Rights and Freedoms [by s. 32(1)] "is confined to government action," whether by the federal, provincial, or territorial governments, "to protect individuals against the coercive power of the state," and therefore its provisions cannot be applied to institutions in the private sector. Second, Ontario universities, even though constituted by provincial acts of parliament, and even though heavily funded by governments to serve and to carry out government-mandated policies in education, are nevertheless "legally autonomous" institutions that "do not form part of the government apparatus" and are thus "private entities."[16] Third, even if these universities' mandatory retirement policies do violate section 15 of the federal Charter of Rights and Freedoms, as does section 9(a) of the Ontario Human Rights Code, nevertheless the raisons d'être for mandatory retirement policies are those that fully meet the test of section 1 of the Charter, namely that "it guarantees the rights and freedoms set out in it subject only to such reasonable limits prescribed by law as can be demonstrably justified in a free and democratic society."[17]

Justice Gérard La Forest wrote this majority report, on behalf of himself, Chief Justice Dickson, and Justices Sopinka, Gonthier, and Cory; the dissenters were the two female Justices, Bertha Wilson and Claire L'Heureux-Dubé. Their arguments for and against contractual mandatory retirement will serve well in elucidating why the administrations of virtually all Ontario universities still seek to maintain mandatory retirement. There are five key arguments used to support mandatory retirement, the first three of which figured strongly in the Supreme Court decision. The remainder of this study will review and analyze these.

(1) To Open Employment and Promotion Opportunities for Younger Workers
Kesselman notes that this is one of the most commonly cited arguments for mandatory retirement.[18] As Justice La Forest himself stated: "[T]he problem of unemployment would be aggravated if employers were unable to retire their long-term workers"; and, further, that "[M]andatory retirement has become part of the very fabric of the organization of the labour market in this country." For La Forest, this situation fully justified the restrictive clause s. 9(a) in the Ontario Human Rights Code, which the dissenting Justice Claire L'Heureux Dubé called "blatantly discriminatory" (indeed the whole Code). Justice La Forest's chief point was that since Canadian universities operate "as a closed system with limited resources ... there is a significant correlation between those who retire and those who may be hired." If mandatory retirement were to be abolished, he contended, "the young [prospective faculty members] would suffer"; and that situation, furthermore, would deprive university students "of younger faculty members and of the better mix of young and old that is a desirable feature of a teaching staff," gratuitously adding, without citing any proof, that "there is at present a significant problem of an older teaching staff in universities."

Evidently he did not consider this to be serious problem in the Canadian judiciary, a system in which federally appointed judges may continue to serve until seventy-five. Does the Canadian judiciary constitute less of a "closed system" than do our universities? Just what is meant by a "closed system," especially when Canadian universities recruit faculty from across the world (as the Canadian judiciary does not)? La Forest also stated, in making these arguments, that "I am not suggesting that discrimination against the old is as such justifiable to alleviate the difficulties faced by the young."[19] But surely that is precisely what he was advocating: age discrimination, in order to alleviate purported and unsubstantiated difficulties that young academics with PhD degrees face in securing university appointments.

Kesselman however, attacks the view that contractual mandatory retirement is required to increase the employment of "younger workers," by contending that the arguments adduced to support contractual mandatory retirement run "counter to elementary economic principles," and constitute what economists calls the "lump of labour fallacy": namely, that at any specified time the "economy offers only a given total amount of work."[20] In a remarkable study published by the Canadian Labour Congress—still obdurate in defending mandatory retirement—its research staff contended that any defence of compulsory retirement on the grounds that it creates new jobs for the young is both "bad economics and dangerous ethics."[21]

In any event, economic history also invalidates the "lump of labour fallacy." Thus, the Canadian economy, with a continuously rising population, having grown from 18.224 million in 1961 to 32.040 million in 2004,[22] provides an ever increasing demand for skilled and educated labour—and that certainly includes university professors, especially since student enrolments have grown so much more rapidly than has the number of available professors. We are told that Ontario alone will suffer a net deficit of some 10,000 professors or more over the next fifteen years. As Justice Bertha Wilson observed in *McKinney v. University of Guelph,* any "exclusion" of younger academics from universities, if it exists, "flows solely from the government's policy of fiscal restraint";' and thus it has nothing to do with the continued employment of senior faculty.[23]

At the same time, of course, nobody would contest the view that universities, along with similar institutions, need to be continuously rejuvenated with "new blood"—younger people who may have newer, fresher, or at least different ideas and educational experiences, if not necessarily better ones. But to suggest that universities can hire new professors only by getting rid of older professors, by compulsory retirement is, on the face of it, absurd for reasons beyond those of the "lump of labour fallacy." For many university professors leave for reasons other than mandatory retirement: some do choose voluntarily to retire before sixty-five, and undoubtedly many more leave to seek better or better-paid opportunities elsewhere (especially in the U.S.—and increasingly even in the United Kingdom). For example, in 2003, the Department of Economics at the University of British Columbia hired four new professors, not all of whom were young (one older professor was lured away from the University of Toronto's Economics department); but it also lost four professors, none of whom retired, and thus all of whom found better opportunities elsewhere. Obviously, even within a "static state" employment economy a university can receive "new blood."

This argument to justify contractual mandatory retirement also implies that, if professors were not forced to retire at sixty-five, they would stay on forever: "Old professors never die; they just fade away" (to misquote General Douglas MacArthur's famous farewell speech). Professors do die, before and after sixty-five. Furthermore, most of those who might choose to continue after sixty-five would probably wish to retire in a very few years, at 68 or 70; very few would stay on into their seventies. That would indeed be "a few of the very few," because the historical experience of Manitoba (from 1982 to 1996) and Quebec (from 1983) has been that most professors retire around or indeed before the age of sixty-five. Currently in Quebec, the average age of retirement, in all universities, is 63.5.[24] Finally, Kesselman cites a study undertaken years after Manitoba and Quebec banned contractual mandatory retirement, which found "the effects on labour force participation rates of people sixty-five and over to be statistically insignificant."[25]

The most recent attack on the "lump of labour fallacy", fully supporting Kesselman's views with far more extensive data, appears in a recent article on "Mandatory Retirement and Older Worker Employment," in which the authors, Shannon and Grierson, analyzed employment statistics for the two provinces without mandatory retirement, i.e., Quebec and Manitoba (the latter, 1982 to 1996). On that basis, they contend that the numbers who chose to remain in the labour force after sixty-five are very small and that elsewhere no statistical support can be adduced for the still common and indeed prevalent view that "eliminating mandatory retirement will significantly worsen the job prospects of younger workers or substantially boost average labour costs by forcing employers to retain large numbers of better-paid older workers."[26]

Exactly comparable data for the U.S. for which the age of mandatory retirement had been 70, from 1978 to 1993, are not yet available. But a recent economic analysis of the effects of abolishing contractual mandatory retirement in 1994 shows that retirement rates before and after 1994 were "very similar," so that "none of the differences ... is even close to [having] statistical significance."[27] The only significant difference, and one to be expected, was in the proportion of the faculty who voluntarily chose to retire, or chose not to, at the ages seventy and seventy-one, the former age of mandatory retirement. If some differences in pension plans (private and state) within the US and between the U.S. and Canada may slightly skew results, the statistical conclusion remains inescapable that, even if the "lump of labour fallacy" were true for universities, the impact that terminating mandatory retirement has had upon the employment of the young is inconsequential.[28]

Finally, if the young, or anybody else, face employment constraints, the forced retirements of all over sixty-five would be no more justifiable than excluding women from such jobs on the specious and indeed odious grounds, so commonly enunciated many decades ago, that their employment would deny employment income to male "breadwinners" (i.e., responsible for feeding a "large family"). Is age discrimination—for that is the core of mandatory retirement—today any more justifiable, and any less despicable, than sexual (gender) discrimination?

(2) To Obviate the Need to Monitor Employment Performance, and thus to Maintain Tenure, with Perceived Productivity Problems in Aging Professors

A closely related argument for allowing contractual mandatory retirement is that productivity declines as employees approach age sixty-five, or that their compensation rises faster than does their productivity, so that contractual mandatory retirement permits employers to discharge such unproductive or costly workers gracefully and equitably at sixty-five, without requiring costly or harsh performance monitoring.[29] In *McKinney v. University of Guelph,* Justice La Forest, asserted, with virtually no documentation, "that on average there is a decline in intellectual ability from the age of 60 onwards,"[30] so that "to raise the retirement age [beyond sixty-five], then, might give rise to greater demands for demeaning tests for those between the ages of 60 and 65." In citing publications of two University of Toronto economists, Morley Gunderson and James Pesando, La Forest painted an even more dire picture of the likely consequences of abolishing contractual mandatory retirement at sixty-five:[31]

> [D]ismissals of older workers would likely increase; monitoring and evaluation of all workers would also increase; so too would continuous monitoring and evaluation; ultimately, compensation of older workers would fall and that of younger workers would rise; [and] the importance of seniority would be affected.

As Kesselman acidly comments, "[I]t is ironic that the mean age of the justices deciding the case was 65; three were over sixty-five, and Supreme Court justices can continue holding office until 75."[32] Justice L'Heureux-Dubé made the same observation in her rebuttal, and noted that American university professors were not then required to retire before seventy.[33] Why should Canadian academics be forced to retire so much earlier? Justice Wilson asked an even more pertinent question:[34]

> Is the mandatory retirement policy a reflection of the stereotype of old age? Is there an element of human dignity at issue? Are academics being required to retire at age 65 on the unarticulated premise that with age comes increasing incompetence and decreasing intellectual capacity? I think [that] the answer to these questions is clearly yes and that s. 15 [of the Charter] is accordingly infringed.

Is there any evidence to support the view that academic ability or performance declines with age? In recent studies and in his chapter in this volume, Kesselman denies that there is any such convincing evidence, for "even if an individual's work skills do eventually decline with advanced age, there is no evidence that this occurs abruptly at sixty-five or as early as sixty-five in most occupations." He also notes that in many so-called "white collar" occupations, according to one major study, "reliability and especially experience seemed to compensate for the effects of somewhat reduced physical abilities."[35] He contends that those "who do experience declining physical, sensory, or mental faculties are more likely to retire voluntarily prior to age 65," so that an "individual's labour force participation and retirement decisions can be modeled as a utility-maximizing choice." In other words, any problems of declining performance are largely a self-correcting and self-adjusting phenomenon: that "most workers who find their productivity declining tend to self-select into early retirement."[36]

In my own view, many academics—though generally more so in the humanities than in the pure sciences—prove to be more productive in their later than in their earlier years, because they then possess a far greater stock of intellectual capital on which to draw, from accumulated research over so many years, and because they have better mastered the art of academic writing after so many years of experience.[37] Some of these are the ones more likely to wish to continue their academic career beyond that arbitrary retirement age of sixty-five.

Support for this argument may be found in the previously cited article by Ashenfelter and Card. Their multi-variable regression analyses of retirements in American universities in the 1990s provide two very interesting results. First, the statistical analyses "suggest that salary exerts a strong negative effect on the probability of retirement": i.e., that the higher the relative salary the lower is the likelihood of choosing retirement; while, on the other hand, "pension wealth works in the opposite direction, but has a considerably smaller effect." That is hardly surprising, since salary increases, combined with years of service, largely determine the amount of pension received at retirement. One may also assume that salary levels, especially at private universities, are very strongly correlated, *ceteris paribus,* with academic merit and productivity, especially in terms of peer-reviewed articles and monographs.

Even more interesting are the statistical comparisons between the major research institutions, chiefly private, on the one hand and public research and non-research institutions on the other. In the former, which includes the famous Ivy League universities (but also Berkeley), only 3 percent of sixty-year-old faculty were still employed at age seventy-three, when mandatory retirement was in force [up to 1993]; but after its abolition, that proportion has risen to "30 percent or even higher." In the other public and non-research institutions, however, "the expected fraction of sixty-year-olds who remain at work until seventy-three has risen to [just] 10 percent."[38] That, of course, corresponds to the first result, indicating a negative correlation between salary levels—obviously far higher (and well-documented) salaries at the very best American universities—and optional retirement.

In view of the academic achievements of so many senior faculty at these universities, with so many Nobel prizes, can one doubt society's gains by allowing such professors to continue research and teaching into their early seventies? Indeed, in Kesselman's chapter 9 can be found one the most powerful arguments against mandatory retirement: an extract from the editorial on *The Process of Aging* in Harvard's student newspaper, the *Crimson* (23 February 2004). It is a paean of praise to older professors (i.e., those over seventy), in particular contending that their contributions in both research and teaching provide "one of the most exciting and unique aspects of being a Harvard student" so that "we are grateful to these professors for dedicating their lives to academia."[39]

Over the past fifteen years or so, my conversations with various department chairs and deans at the University of Toronto have led me to believe that most academic administrators hold instead strongly contrary views: specifically that academic performance generally does decline with age, from the early sixties; that underachievers would not voluntarily resign at sixty-five, if mandatory retirement were abolished; and thus that, if such professors did not do so, administrators would be faced with the invidious task of monitoring their performance and then of terminating the employment of those found truly wanting.

In rendering the majority decision in *McKinney v. University of Guelph*,

Justice La Forest certainly took such a view: that the costs and social unpleasantness involved in terminating the employment of unproductive professors "for cause" were greater than the costs incurred in losing good professors. His chief argument was that mandatory retirement was a necessary "quid pro quo for a tenure system with minimal peer evaluation," whose maintenance was to be justified in protecting academic freedom and thus "in enabling universities to be centres of excellence on the cutting edge of new discoveries and ideas." That view, in turn, is linked to a related argument cited earlier: namely, that mandatory retirement "ensures a continuing, and necessary, infusion of new people." In La Forest's view this laudable objective, sought within "a closed system with limited resources"— that is, in all Canadian universities— "can only be achieved by departures of other people." Therefore, if mandatory retirement were to be abolished, university administrators would be forced to impose or ensure a sufficient number of such "departures" by measures that would likely lead to the abolition of tenure. In his alarmist view, universities would have to institute "a stricter performance appraisal system" and one "likely requiring competency hearings and dismissal for cause," which, furthermore, "would be difficult and costly and constitute a demeaning affront to individual dignity."[40]

In a statement reported recently in *University Affairs,* in December 2003, Professor David Foot reiterated what is undoubtedly still a very common view, in stating that "there will be no more coasting to retirement," if professors were no longer required to leave at sixty-five, and, furthermore, that an end to mandatory retirement would "undermine the purpose of tenure, which is supposed to free professors from excessive supervision to ensure academic freedom."[41]

In her reply to Justice La Forest, Justice L'Heureux-Dubé rejected the "proposition that abolition of mandatory retirement of university faculty and librarians would threaten tenure as a result of increased performance evaluations," while also challenging Justice La Forest's argument that an evaluation scheme will "constitute a demeaning affront to individual dignity." Of her many profound observations, the most important was that, "those jurisdictions which have eliminated mandatory retirement of university faculty or librarians have not experienced any increase in so-called destructive performance evaluations, or any infringement of academic freedom," so that the "tenure system remains firmly in place." Noting that, by 1990, 15 percent of American universities had voluntarily eliminated provisions for compulsory retirement, she stated that "not a single university has abolished tenure," in either the U.S. or Quebec.[42] Her statement remains just as true today, more than ten years after contractual mandatory retirement was fully abolished in the U.S. Today most American universities continue to thrive, with, on average, much higher salaries and research grants than are enjoyed by their Canadian counterparts.[43]

Needless to say—yet it must be stated clearly—any provincial legislation to abolish contractual mandatory retirement in Canadian universities cannot and may not permit such universities to abolish tenure from the age of sixty-five. If that were permitted, Canadian university administrators could then do "an end run" to achieve their former goal of dispensing with professors at that age by engaging in arbitrary dismissals, i.e., without specifying and documenting the legitimate reasons for doing so.

Let us be clear on the true meaning of the term tenure, as it applies now to North American universities. It does not mean—contrary to popular opinion—guaranteed university employment. Rather, it means that a university is not permitted to engage in such arbitrary and capricious dismissals. In 1967, the Board of Governors of the University of Toronto agreed to adopt tenure, by legally defining it as "a continuing full-time appointment which the University has relinquished the freedom to terminate before the normal age of retirement except for cause," and "after following certain procedures," involving "due process."[44]

Any such dismissals would be based on the evidence documented in the annual reports that, in all major North American universities, each departmental chair, institute director, and/or faculty dean is required to produce for each faculty member within his/her jurisdiction. The chief purpose of those reports is, of course, to permit a reasonably objective assessment of the "merit award" or the "Progress Through the Ranks" (PTR) component, which is added to any negotiated Cost of Living Allowance (COLA) increase, in determining faculty salaries for the following year. The departmental chair's assessment, often undertaken in collaboration with the associate chairs, is based on two documents, produced each spring: (1) the student evaluations of the professors' courses (undergraduate and graduate); and (2) the *Annual Activity Report* that all faculty members are required to file (or else forgo any merit award increase), to provide detailed evidence on their publications (refereed and non-refereed), those either accepted for publication or currently in press, conference papers and public lectures, and research in progress (with explanations on how the research has been conducted). This very detailed report also documents the professors' "service contributions" to the university and the community. The merit award (PTR) is thus based on the accumulated points produced by some weighting of the three components: teaching, research, and academic service.[45] As Justice L'Heureux-Dubé herself observed, that is precisely the form of almost universal "monitoring" that critics so fallaciously contend would be too costly or socially painful to administer.[46]

Although some student course evaluations may be unfair, since some students, guaranteed anonymity, may be vindictive for unjust reasons, these evaluations combined with the faculty's annual activity reports have provided a very valuable tool for faculty "monitoring" over the forty or more years that they have been in use. At the University of Toronto, those in the senior salary category receive an annual salary increase, if any, based *solely* on "merit awards," determined by this very same process. Furthermore, it has become an accepted custom that any tenured faculty member who does not earn any merit award increase for three consecutive years may be subject to "dismissal for cause."

To be sure, undertaking the legal procedures to dismiss an incompetent professor for cause may be costly and unpleasant. Nevertheless, the potential cost involved hardly constitutes a significant argument for retaining mandatory retirement, because the occasions requiring such dismissals would still be few and far between. First, many Ontario universities (e.g., the University of Toronto) have been quite successful in convincing such "undesirable" professors to retire early, even if the solution is more often a "buyout" package than a threat of "dismissal for cause." Indeed, many American colleagues have told me that moral suasion and attractive "buyout" packages are the common remedies that their university administrators employ to get rid of unproductive faculty (including those under seventy); and they scoff at the notion that abolishing mandatory retirement has burdened them with supposed "deadwood." Second, we must recall Kesselman's arguments and evidence about "self-selection": i.e., that the vast majority of such professors, who "could not make the grade," are much less likely to continue teaching as full-time academics, after the normal age of retirement.[47] Third, since the 1970s, the monitoring procedures undertaken by formal departmental academic committees engaged in hiring (involving "job paper" seminars), and then in promoting colleagues to tenure and finally to Full Professor, are now so rigorous that not that much "deadwood" survives, certainly not at the major Canadian universities. Is it likely that, in the absence of mandatory retirement, the experience of Ontario universities would differ substantially from that of universities in Quebec or the U.S.?

In my view, Ontario university administrations are guilty of crass hypocrisy in contending that they require mandatory requirement in order to get rid of academic "deadwood," relatively "painlessly," and no later than the age of sixty-five, because, in my experience (i.e., in the Economics department at the University of Toronto), no retired colleague who wishes to teach a course has ever been refused, whatever has been his/her academic record in teaching and research. The word "deadwood" never, ever arises in such cases, not even in the case of many retired colleagues now in their seventies who continue with some stipendiary teaching. If the university really upholds contractual mandatory retirement for this specious reason, then it should be more selective in hiring stipendiary teachers, who, after all, have no basic contractual rights to teach or to continue teaching. But in fact, our department and so many others are so desperately short of teachers or lecturers to hold necessary classes that virtually no offers to teach a course will be rejected. In our department, the proportion of courses given by stipendiary lecturers, including retirees, ranges from

35 to 40 per cent. The only constraint is the supply of funds to pay what are, in fact, miserly stipends (about $5,000 per semester course).

(3) Mandatory Retirement as a Contractual Agreement

As noted earlier, the proper term to be employed is "contractual mandatory retirement"; and my colleagues Gunderson and Pesando believe that mandatory retirement at sixty-five is socially justifiable on the grounds that it is a feature of so many freely negotiated labour contracts.[48] Their arguments, or those published in 1988, deeply influenced the majority decision in both *McKinney* and *Dickason*; and certainly in the former Justice La Forest justified contractual mandatory retirement several times *inter alia* on these very grounds.[49]

In the case of the University of Toronto itself, this argument is invalid in light of the university's history. According to its historian, Martin Friedland, the retirement benefits that the administration secured from the Carnegie Corporation for academics "at non-sectarian institutions," on the eve of World War I were not available until a professor had reached the age of seventy; and he assumes that most did continue teaching until that age. If, following the Great War, sixty-five became the customary age of retirement, university administrations nevertheless granted an exemption to most faculty members who wished to continue teaching, with full salary and benefits, to sixty-eight, or sometimes to seventy.[50] In 1955, President Sidney Smith raised the "official" age of retirement to sixty-eight, in effect making it mandatory, while still permitting voluntary retirement with full benefits at sixty-five. That mandatory retirement age remained unchanged for the next seventeen years, until 1972, "when it was abruptly lowered to sixty-five … without consultation with UTFA [University of Toronto Faculty Association]," as stated in UTFA's official history.[51] John Evans had become the new President on July 1, 1972, and the new Governing Council, marking a radical reorganization of university government, held its first meeting on July 4.[52]

That unilateral action, by administrative fiat, took place five years before the faculty had finally achieved sufficient organizational cohesion and power to gain rights of collective bargaining, in 1977, through the Memorandum of Agreement. The administration finally and most reluctantly agreed to sign this document only after the faculty had twice voted to consider union certification as the only effective alternative. The university administration was thus convinced that such certification would inevitably lead to faculty strikes or other serious disruptions to academic life. As William Nelson comments, in his history of UTFA, on the university's imposition of mandatory retirement in 1972: a "few years later the 'frozen policies' clause in the Memorandum would have made such a unilateral change impossible"—i.e., the clause stipulating that university policies and traditions in force at the time the Memorandum was signed could henceforth be changed only by mutual consent, through collective bargaining.[53] It should also be noted that the Memorandum of Agreement does not permit the University of Toronto Faculty Association to go on strike, but, by 1982, it had been revised—again under the threat of full union certification and almost certainly a strike—to provide various alternative measures of mediation and arbitration (with de facto compulsory arbitration, as the last resort).[54]

Therefore, as far as the University of Toronto itself is concerned, the argument to justify contractual mandatory retirement at sixty-five on the basis of "freely negotiated contracts" certainly does not apply. Consider these two scenarios, the first a "counter-factual." Suppose that the University of Toronto had not (in 1972) imposed mandatory retirement at age sixty-five, and subsequently, after the adoption of the Memorandum of Agreement, suppose that it had sought to do so. The executive and bargaining committee of UTFA would have responded by pointing to the "frozen policies" clause of the Memorandum and then would have stated that this was not an issue for negotiation. This is not idle speculation, because in 1985, under the leadership of and at the urging of then President Michael Finlayson, the UTFA council endorsed the current resolutions of the Canadian Association of University Teachers (CAUT) condemning contractual mandatory retirement and it then passed "a resolution opposing mandatory retirement and urging a flexible retirement policy on the administration."[55] Those resolutions have been endorsed by many subsequent UTFA annual general meetings, most recently on April 15, 2004.

Consider the opposite scenario. Suppose that, some time after 1985, the UTFA executive had sought to bargain with the university to abolish contractual mandatory retirement, in compliance with the Finlayson resolution and those of subsequent UTFA annual general meetings. The administration similarly would have pointed to the "frozen policies" clause of the Memorandum of Agreement and retorted (as it has often done, in effect) that the issue was and is not subject to negotiation.[56] But suppose, further, that the administration would have been willing to negotiate this issue. What costly (and demeaning) concessions would it have demanded in return? Of course, for reasons cited earlier, an abolition of mandatory retirement combined with the abolition of tenure would be completely unacceptable.

In the light of this evidence, it is all the more amazing that Justice La Forest, and those supporting the majority decision in *McKinney*, subscribed to and so enthusiastically endorsed these fallacious arguments about "freely negotiated contracts" that supposedly permitted contractual mandatory retirement in Ontario universities. Not only had the CAUT and then UTFA (from 1985) adopted resolutions opposing mandatory retirement, as contrary to the Canadian Charter of Rights and Freedoms, but both had supported the appellants in that Supreme Court case (and the earlier court case heard in the Ontario Court of Appeal).[57] Furthermore, one of the listed appellants in *McKinney* is the York University Faculty Association.[58] Did the justices really believe that faculty associations in Ontario universities were somehow, and improperly, reneging on "freely negotiated contracts?" Or did they simply ignore the faculty views and their published resolutions? Let us remember that this celebrated case concerned, and concerned only, Ontario universities. Surprisingly, even the two dissenting justices, L'Heureux-Dubé and Wilson, did not comment on this vital issue.

One may concede, however, that in many other cases, particularly those concerning industrial workers such as the United Auto Workers, union contracts quite clearly do contain freely negotiated clauses to permit mandatory retirement. In 1980, the Canadian Labour Congress adopted a resolution, valid to this day, which explicitly stated that "the organized labour movement has fought hard and long legislative battles to establish the mandatory retirement age of sixty-five

(65) years"; and therefore it "resolved that the Canadian Labour Congress oppose the erosion of the mandatory retirement system."[59]

Nevertheless, one may question whether two contracting parties, the employer and the union of employees, have the moral right to abrogate what is clearly a minority right, namely a freedom from age discrimination. In discussing labour union contracts in her dissent to *McKinney*, Justice Wilson posed just that question: "whether citizens can contract out of their equality rights under s[ection] 15" of the Charter. She had "serious reservations that they can be [so] contracted out of," since "equality rights lie at the very heart of the Charter."[60]

An opposing view, one argued by Gunderson and Hyatt, is that such age discrimination in employment can be justified on two linked grounds.[61] First, it differs from "other enumerated grounds in human rights codes such as sex or visible minority status, in that all individuals can expect to reach age sixty-five (with good fortune), but all persons cannot expect to be female or a visible minority." Second, we are entitled to bargain away certain rights and benefits that might accrue to us later, if we did survive to sixty-five, in return for superior current benefits, so that "then presumably we are discriminating against *ourselves*—or at least ourselves in the future." Both arguments must be contested, despite some dubious support to be found in Justice La Forest's majority decision—though with arguments subsequently contested by Justice L'Heureux-Dubé.[62] Section 15 of the Charter does not evaluate the conditions of discrimination. The ones cited are all completely and equally invalid. Otherwise, we would find ourselves in an alien world akin to George Orwell's *Animal Farm,* in which "all animals are equal, but some animals are more equal than others."[63] Furthermore, if one were to choose to alter one's sex (as some certainly do, by physical operations) or to change one's religion—thus removing the "permanence" of such conditions—would any court seriously consider this to be a valid argument to permit sexual (gender) or religious discrimination?

Second, members of a labour union, in agreeing to a contract that stipulates mandatory retirement for those over sixty-five, are certainly not, in most cases, making a conscious decision to weigh their own individual present benefits over their own individual future benefits (i.e., by trading away their own individual right to continue working after sixty-five), even if most probably do have a high discount rate, and live primarily "for the moment." Instead, most are essentially bargaining away the rights of a few current older workers to improve their own personal benefits—and that certainly is age discrimination.

The Gunderson-Hyatt argument fails, as Kesselman observes in Chapter 9, to distinguish between an individual contract, freely negotiated by that one individual with his/her employer, and a collective union agreement. The distinction is important, because in Canada contractual mandatory retirement "is highly concentrated in work covered by collective agreements ... not individual contracts." One may also add that even if rank-and-file younger union workers might well accept the common argument of their union leaders to justify mandatory retirement—that it is a necessary trade-off to ensure the receipt of their pension benefits at sixty-five, an important issue next to be addressed—most such contacts are written by their union leaders (along, of course, with their employers) and not by the rank and file. Furthermore, as Kesselman also observes, "even if all CMR [contractual mandatory retirement agreements] were clearly based on consensual agreements between individuals and their employers, one might question the ability of most people to predict their situation and needs many years into the future."[64]

(4) Promoting the University's Goal of Greater Diversity

This is an argument that the Canadian Supreme Court (in 1990 and 1992) did not consider, but it is certainly one that had been maintained by the University of Toronto.[65] As noted earlier, it was also the most serious consideration in exempting university faculty from the otherwise complete U.S. ban on mandatory retirement in 1986 (in the Age Discrimination in Employment Act).[66] Certainly increasing the diversity of the university's faculty is a laudable goal—if universities could agree on what is meant by "diversity." Yet it seems invidious, and contrary to the university's intellectual traditions, to manipulate both retirements and new employments to ensure that older males are replaced by females and/or other males of non-European origin. We may well ask: does the university really mean that, faced with scarcities of professors in certain academic fields, it will allow considerations of supposed diversity to supersede considerations of intellectual merit, talent, and experience?[67]

In any event, as a device to achieve such laudable goals of greater gender (and ethnic) diversity in university faculties, mandatory retirement is no more effective or necessary than it would be as the only means of hiring "new blood," young or old, for reasons that were argued above. More important, it is unethical: for one cannot condone a blatant form of age-discrimination in order to combat the perceived ills of a heritage of another form of discrimination.

Unfortunately, these views might not be supported by some Canadian jurists, because section 15(2) of the Charter does explicitly permit age and other forms of discrimination to help achieve this goal (known as affirmative action), in stipulating, in the 1985 amendment, that "Subsection (1) does not preclude any law, program or activity that has as its object the amelioration of conditions of disadvantaged individuals or groups including those that are disadvantaged because of race, national or ethnic origin, colour, religion, sex, age or mental or physical disability."[68] In *McKinney,* the Supreme Court agreed not to invoke this clause on the grounds "that younger academics do not constitute a 'vulnerable' group within the meaning of the case law."[69] But it is significant the Supreme Court also did not refer to the arguments used to obtain that exemption from contractual mandatory retirement for American professors in the Age Discrimination in Employment Act (1986 amendment), which, as noted, Congress permitted to expire on December 31, 1993.[70]

(5) Mandatory Retirement as a Fiscal Necessity

When, in 1972, the University of Toronto administration arbitrarily imposed mandatory retirement, at the reduced age of sixty-five, its ostensible reason was to permit the university to cope with its current financial stringency, which, of course, has always been "dire." The university's fiscal rationale was as follows: that, for every two full professors who retired at sixty-five, the university could either hire, as replacements, three junior assistant professors, for the same price; or it could reduce the university's aggregate salary budget by hiring only two to replace those two forced to retire.[71] The blame for this fiscal crisis was laid—then as now—on the Ontario provincial government, which has consistently ranked last, or less frequently, next to last, amongst Canadian provinces in per capita university funding, in the past thirty years.[72] Nevertheless, as noted earlier, all Canadian university administrations, in all provinces but Quebec, from the very best to worst funded, still wish to retain mandatory retirement, for whatever reasons they deem best at the time.

Does the University of Toronto still reap such financial gains from imposing mandatory retirement at sixty-five? In our Economics department, there do not appear to be any such gains, not for at least five years. Consider the arithmetic: a newly hired, freshly minted PhD, with absolutely no teaching or other academic experience, can now expect to receive a salary about 75 percent of the average final salary for a retiring professor; but for the first five years or so he/she will do only 60 percent of the retired professor's teaching load, i.e., no more than three semester courses, instead of the customary five courses. That is the necessary part of the current market price that we must pay in order to attract new faculty, rather than losing them to many other universities, especially American, whose standard teaching load is just three or four semester courses. Furthermore, we rarely succeed in filling all new positions in this extremely competitive market, dominated by American universities with no mandatory retirement. Therefore, it does not appear that young economists are being hindered in securing academic employment, or would be if mandatory retirement were abolished.[73] Since many retired colleagues continue to be engaged in fruitful research that produces publications, the university can hardly argue that it is getting a compensatory dividend in the form of published research from the newly hired professor, who is not likely to publish successfully for several years, in many cases.[74]

Is the fiscal argument, therefore, without any merit? No: the university does gain, at least by hiring a retiree, on a stipendiary contract. In my own case, my total stipend for teaching four semester courses is about 15 percent of my former salary. As James Turk (Executive Director of the Canadian Association of University Teachers) states so succinctly, in the December 2003 issue of *University Affairs:* "It is not that they [Canadian universities] don't want older teachers to teach, it's that they don't want to pay them as much to do so."[75] In other words, contractual mandatory retirement is a useful device—and from the university's point of view, a necessary device—to permit it to offer more courses by so cheaply employing retirees (and other stipendiary lecturers). Of course such stipendiary teaching, undertaken quite voluntarily, does not constitute exploitation, an unduly charged word that we economists rarely use. If one has taught here for thirty-five years before retirement, as I had done, the pension income is now sufficiently good enough that most currently retiring faculty in my situation would not require such stipendiary teaching to maintain their standard of living.

On the other hand, mandatory retirement in most departments at the University of Toronto does mean the loss of one's private office. The price to be paid to obtain some semi-private space, in the form of a lockable cubicle in a retiree's room, is to agree to provide such teaching. That office space, though semi-open and lacking in privacy (and peace and quiet), does at least allow such a retiree to have fruitful contacts with departmental colleagues, access to departmental amenities (fax, photocopying, mail services, computer access), and, above all, close access to the Robarts Research Library, the third- or fourth-best university library in North America.

As a recent retiree from the University of Toronto, I and many others can fully appreciate Justice L'Heureux-Dubé's conclusions on the harmful nature of mandatory retirement, even if personally I am less afflicted or disadvantaged than many others, especially those female professors who, in raising their own families, have been unable to serve as many years in university employment as I have. Most eloquently, she contended:[76]

> [that] its negative effects significantly outweigh any alleged benefit associated with its continuation. Mandatory retirement arbitrarily removes an individual from his or her active worklife, and source of revenue, regardless of his or her actual mental or physical capacity, financial wherewithal, years of employment in the work force, or individual preferences. The continued opportunity to work provides many individuals with a sense of worth

> and achievement, as well as a source of social status, prestige, and meaningful social contact; and on the evidence, there is no basis for denying to a segment of the population, i.e., those aged 65 and over, the protection of legislation which is of fundamental importance in the area of employment discrimination.

Final Considerations: The Link between Normal and Mandatory Retirement

A vitally important question must now be addressed: is there a difference between the concerns of university professors and, say, unionized industrial workers about mandatory retirement? On the basis of a computerized Google search, Professor John Myles concluded that concerns about mandatory retirement seem "to focus on two occupational groups: university professors and judges"—two groups whose members disproportionately continue to enjoy and relish their occupations into and sometimes past their sixties.[77] In many other occupations, however, especially in industrial occupations governed by labour union contracts, a substantial majority of workers do not enjoy their jobs, especially if they are physically arduous, laborious, and tedious. For so many unionized workers, their goal is to retire earlier than at the age of sixty-five, and certainly not later.

For this reason any faculty association or other advocacy group in Canada that seeks to eliminate mandatory retirement (outside Quebec) must strictly beware engaging in any form of linkage between the abolition of contractual mandatory retirement and the age of normal retirement. We must therefore respect the views and concerns of labour union leaders and clearly understand that the primary reason why they are so opposed to the abolition of contractual mandatory retirement is the fear, rational or not, that in doing so the ultimate consequence may be an increase in the age of retirement to sixty-eight or seventy. Any threat to change the age of voluntary retirement, and thus any threat to the aspirations of the vast majority of wage-earning employees to secure their full pension and other retirement benefits at sixty-five, is certainly bound to harm the public case for abolishing mandatory retirement. There must be full recognition that only a few will freely choose to continue with their employment past sixty-five, and that society stands to gain more than it will lose from such choices.

Such a fear, entertained by union leaders (and evidently inspired by employers), might have seemed rational when the Canadian Labour Congress first enunciated it in 1980, and thus before mandatory retirement was abolished anywhere in Canada or the United States.[78] But given the now long historical experience with the abolition of mandatory retirement, especially outside the academic world—in Manitoba since 1982, in Quebec since 1983, and in the U.S. since 1986 (university professors from 1994)—do such fears now have any rational foundations? Nowhere in any of these jurisdictions have employees, whether or not unionized, yet lost their right to enjoy their full pension and other retirement benefits at sixty-five (or even earlier, though with somewhat reduced benefits), *if they have freely chosen to retire at that age.* Yet one cannot dismiss the fear that this situation could change in the future, especially since so many Canadian employers have encouraged the view that, if forced to abolish mandatory retirement, they would seek to raise the age at which retirement benefits would be provided to sixty-eight or seventy. In the United States, legislative measures have been initiated to increase the age to receive pensions and other retirement benefits to sixty-eight (in stages, by age cohorts), even if for reasons unrelated to mandatory retirement.

One closely related aspect of very serious concern that I find in Kesselman's otherwise excellent papers on mandatory retirement is the implication that society would gain economically from abolishing contractual mandatory retirement, if that meant that those who continued to be employed would not only pay more income and payroll taxes, but would also reduce the burden on public pensions by delaying their initial payments from such pension schemes. Therefore—and this is by far the more ominous and least desired implication of his publications—society would gain all the more, if the normal or standard age of retirement were raised from sixty-five, to say, sixty-eight or seventy.[79] Whatever the long-term economic merits and justifications for such arguments may be, North American society today still holds as sacrosanct the right to retire with full benefits at sixty-five.

Will universities really suffer from the abolition of contractual mandatory retirement? Surely, if proof of the pudding is in the eating, we must cite the historical experiences in the United States and

Quebec. Can anyone now make a rational case to demonstrate that universities in Quebec and in the United States have suffered significant costs or losses from the abolition of contractual mandatory retirement? Until someone does so, then the cant emanating from Ontario university administrations should be ignored—or better, should be contested and refuted.[80] Perhaps some may contend that universities in Quebec and the United States have achieved that lower average of retirement by offering financial inducements—but the University of Toronto has, in the past decade, similarly provided such inducements for early retirements, for various reasons (chiefly fiscal, but also including a desire to reduce the number of less-desirable faculty, especially those who received de facto tenure before the 1970s). No one has yet, to my knowledge, made a convincing case that the abolition of contractual mandatory retirement in these jurisdictions has produced a net financial burden on their universities.

What are the gains, for universities, in abolishing contractual mandatory retirement? First, Canadian universities (outside of Quebec) would gain by drawing upon a larger pool of exceptional talent at a just price, so to speak, without taking advantage of those few professors of talent who do wish to continue full-time with their academic careers after sixty-five, as stipendiary lecturers. Second, Canadian universities, in depending so heavily on stipendiary lecturers, may find a significant change: that those who would continue teaching, without mandatory retirement, would do so with a greatly enhanced sense of self-esteem, morale, and enthusiasm—that is, with potentially greater productivity.

Third, universities would find they would lose fewer highly talented professors before the age of normal retirement. Certainly many have left the University of Toronto for universities in Quebec and especially in the United States, in order to escape mandatory retirement, and not just to seek higher salaries (somewhat offset by higher US medical costs) and research grants. Others have left to establish elsewhere in Canada alternative careers that will offer them productive and rewarding lives into their seventies.

Fourth, and conversely, therefore, the University of Toronto and other Ontario universities would have far greater success in attracting similar talent from American universities (or from other universities that do not practise mandatory retirement). Those Canadian universities practising mandatory retirement do not have any hope, whatsoever, of hiring professors over the age of fifty from universities elsewhere that do not have contractual mandatory retirement. Even if such professors may think, at the age of fifty or so, that they probably would retire, voluntarily, at sixty-five, they all want the right to choose when they retire.

I certainly do believe that the intellectual costs—the costs in foregone talent—that we incur by practising mandatory retirement are quite staggering. University administrators who comprehend something of this loss are unjustified in merely shrugging their shoulders and muttering that this is the cost that must be borne. It does not have to be so borne.

11

Challenging Mandatory Retirement in Academe: A "Frivolous and Vexatious Complaint"?

Jean McKenzie Leiper

Introduction

This chapter chronicles my experiences as I approached retirement from my tenured position at King's College at the University of Western Ontario (UWO) in June 2003.[1] It also outlines the steps I followed after retaining legal counsel and mounting a formal challenge to the process of mandatory retirement. This very personal account is embedded in a broader context—the growing recognition that mandatory retirement is a discriminatory practice reflecting ageist attitudes. My title refers to the subtlety of discriminatory attitudes, particularly in the use of arcane legal language to dismiss my challenge.

Late in 2003, as I began to mull over ideas for a paper, I became aware of an interesting pattern in the popular media; I knew retirement at age sixty-five was not universally imposed in Canada, but many high-profile individuals seemed to enjoy special privileges where retirement was concerned. On December 13, the front page headline in the *Globe and Mail* read "Martin Pledges New Priorities," and the accompanying colour photograph of Paul Martin showed a smiling new prime minister who had turned sixty-five just prior to assuming office. To his left sat Adrienne Clarkson, Canada's governor-general, age sixty-four, apparently not contemplating impending retirement.[2] The headline on a second section of the paper announced "Nation Builders of 2003": in a full-colour picture, Ontario's chief justice, Roy McMurtry, was flanked by two other lawyers recently chosen as the Canadians who had done the most to shape the country in 2003. Chief Justice McMurtry was seventy-one.[3] In yet another section, veteran movie director Robert Altman peered deep into the camera under the headline "Close Up," still in full stride at age seventy-eight.[4] The following week, Prime Minister Martin appeared again, defending his right to assume office at age sixty-five, declaring that he hoped to have a ten-year career, and voicing strong opposition to mandatory retirement.[5] That same week, Ed Broadbent stepped back onto the public stage by announcing his plans for a political comeback at age sixty-seven, which resulted in his election to the House of Commons in June 2004.[6]

Canada's national newspaper provides a very public forum for people who occupy centre stage in the government, the legal system, and the entertainment world. No one questions their competence or suggests that they should retire at a predetermined age.[7] Nor is mandatory retirement prescribed for senior incumbents in many other public institutions: Canadian senators and Supreme Court justices are allowed to retain their highly paid, responsible positions until age seventy-five and federal civil servants can continue to work well beyond age sixty-five.[8] In fact, wisdom and age appear to go hand-in-hand for many of our most important elders.[9]

Because I had been forced to retire six months earlier, I was puzzled by the differential treatment accorded various groups. I knew that many people are ready to leave their working lives behind at sixty-five or earlier but in my profession—university teaching—some are reluctant to leave behind a lifetime of research and teaching. Why then do we not have the option of continuing our work? In my case, retirement has been particularly punitive, mirroring the experience of many other women in the academy. The following brief history positions my academic career in the context of larger lifetime pursuits, providing a rationale for my efforts to postpone retirement and remain a committed member of my department.[10]

Developing and Maintaining an Academic Career

I graduated from the University of Western Ontario in 1959, married a year later, and followed the expected route for young women: our first child was born in 1961, my husband provided the income, and I stayed home with our growing family for the next twenty years.[11] Social norms dictated that men be breadwinners and women spend their time raising children.[12] This kind of manipulation of public opinion is a powerful tool for excluding some groups from productive work.[13]

In the mid-1970s I began taking some sociology courses, became fascinated by the material, and decided to enrol in a graduate program.[14] In 1986 I finished my PhD and crossed two important milestones: formal graduation and the completion of my major parenting duties. Interestingly, our youngest child's high school commencement ceremony took place on the same day as my convocation. At the age of forty-eight, ready for the labour market, I began the search for an academic appointment with the same enthusiasm that I had devoted to parenting and community work.[15] However, age discrimination is an insidious process, and I discovered that universities were no more tolerant of older employees than were other institutions. I spent the next eight years in an assortment of contractually limited positions, commuting to various universities in the so-called "Golden Horseshoe" region of Ontario.[16]

In January 1994, I willingly accepted the offer of a tenure-track position in statistics and research methods in the sociology department at King's College. On July 1 of that year, surrounded by supportive colleagues, I began the most rewarding stage in my academic career. I became completely immersed in college life, teaching three demanding courses each year, sitting on college and faculty association committees, and developing a research project that was later to receive generous external funding. I had discovered the joy of academic success, and my retirement date of June 30, 2003 seemed a long way off.

Institutional priorities almost always prevail over individual needs, and my case was no exception. Several years before the formal retirement date, my name began to appear on public lists circulated at college meetings, along with the names of other faculty members slated for retirement. The chair of our department announced my upcoming departure well in advance of the event, alerting colleagues to the fact that they should start considering my replacement. The institutional view of my career was plainly at odds with my optimistic plans to continue. Since I had just received a three-year grant from the Social Sciences and Humanities Research Council of Canada, I spent my summers conducting interviews and presenting papers at academic conferences. During the fall and winter terms, I was busy with my statistics classes. At some level, I thought my performance would justify an extended tenure at the College, but the early announcement of my retirement had had the intended effect. Colleagues began to mention it in casual conversations, usually beginning with, "I had no idea you were old enough to retire…." or "It seems like you just got here and now you'll be leaving."[17]

I began to consider ways of avoiding retirement at age sixty-five, adopting a standard response to queries: "I don't want to retire yet. I'm hoping to work out an alternative plan." In the summer of 2001, I spoke to colleagues on the faculty association's pension committee and later, the members of the Terms and Conditions (of Employment) Committee. The existing Terms and Conditions Agreement between the faculty association and the college was due for review, and I was invited to join the Committee to assist with major revisions to the document, particularly the clauses dealing with retirement age. Jumping at the opportunity, I drafted a section entitled "Delayed Retirement" that replicated the wording of an existing section on "Early Retirement." It seemed only logical that if some faculty members could choose early retirement at age fifty-five, then those with projected low pensions should be offered a chance to work beyond age sixty-five. Unfortunately, the Terms and Conditions Agreement is a complex document subject to intense negotiations with the board of governors, so change occurs very slowly. At the time of writing, early in the summer of 2004, the document remains unchanged, waiting for new committee members to report to the faculty association and present proposed changes to the board.

During this time, I also had informal discussions with other members of the faculty association executive and the principal of the college but no formal changes to existing documents were instituted. By mid-2002, I became concerned that I would lose access to the travel money, computer facilities, and library resources necessary for my research. Also, because my projected pension would be very low, I wanted a chance to build a stronger base by remaining in my tenured position after age sixty-five. In the summer of that year, an article by the principal appeared in *The King's Herald*, a magazine published by the alumni office, serving notice that several faculty members would be retiring within the next few years. My name headed the list.

By the beginning of the fall term, the process of hiring my replacement was fully underway, and I finally realized I was powerless in the face of institutional momentum. The next ten months were to be the most trying period of my tenure at the college. Advertisements for my position were published and I was expected to interview candidates for the position. Students began to ask about my impending retirement and faculty members started to drop by my office, mentally measuring it for their furniture. I continued to respond to comments with the claim that I did not want to retire.

The Legal Argument

As the 2003 spring term came to a close and the official retirement date loomed, I realized I had not resolved the issue in my own mind and decided to explore other possible avenues. I met with Janice Rubin, an employment lawyer with a strong interest in age discrimination in the workplace. She outlined a three-stage process that could lay the groundwork for a challenge to section 15 of the Canadian Charter of Rights and Freedoms before the Supreme Court of Canada. My first recourse had to be through the King's College Faculty Association, giving my colleagues an opportunity to launch a formal grievance on my behalf. If the association declined to accept my grievance, the second step would involve an application to the Ontario Human Rights Commission, claiming I had experienced age discrimination. This step would constitute a formality because the Human Rights Commission deals only with people aged eighteen to sixty-five.[18] The third stage of my complaint would be a Divisional Court Judicial Review of the decision made by the Human Rights Commission. Possible strategies lay beyond these three initial steps, but we agreed specific plans would be premature, because a formal appeal to the Supreme Court would be enormously expensive and require backing from a broad range of supporters.[19]

In the meantime, the college made plans to recognize my retirement. When the dean's secretary contacted me in late April to suggest possible dates for a retirement party, I assured her that I appreciated the gesture but definitely did not want one. After some discussion, the dean suggested that, as an alternative, I entertain several close friends from the college community at lunch. Again, I replied that I appreciated the thought but was opposed to any kind of retirement recognition. I still did not see myself as a retiree. These well-meaning gestures strengthened my suspicion that the college, like any other institution, feels a strong obligation to recognize the passing of people from its ranks. The occasion must be marked by a celebration, formally releasing retirees from their duties as they fade from the academic scene.

At about the same time, the principal's secretary called to request my presence at convocation in June, where I would be granted the status of "professor emerita." I decided to refuse this designation, concerned that it would be seen as formal acceptance of my retirement status. On May 22, my lawyer sent a letter to the principal of the college stating that I would postpone the honour and did not intend to retire from the college on the proposed date, June 30, 2003. Citing section 15 of the Charter as evidence that mandatory retirement was discriminatory, we requested that my employment as a tenured professor continue uninterrupted. She copied the letter to the president of the King's College Faculty Association, asking the association to consider a grievance on my behalf.

On June 3, my lawyer received a call from the King's College lawyer indicating that the principal would not comply with my request. A letter confirming this decision arrived a few days later. She received a letter from the outgoing president of the faculty association in which he stated that the new association executive would need time to deal with the request; a final decision would have to wait until the fall term. He also suggested that because the current Terms and Conditions Agreement included a clause supporting mandatory retirement, the faculty association would have no grounds for acting on my behalf.

The whole issue remained unresolved over the summer, but rumours about my case began to fly around the college, and when the fall term began, plans were made to discuss the proposed grievance at the September meeting of the faculty association. A new president had been installed, and the meeting was scheduled for September 24. My lawyer faxed a detailed letter to the association setting out our position, once again citing section 15 of the Charter as applied to the Ontario Human Rights Code as evidence that mandatory retirement was discriminatory:

> It is our view that an arbitrator would have the jurisdiction to make this finding as the right of an employee to equal treatment without discrimination is implicit in each collective agreement. An alleged contravention of an express provision of a collective agreement is not a condition precedent of an arbitrator's authority to enforce substantive rights and obligations of the *Ontario Human Rights Code.*

In the meantime, the president of the faculty association contacted a lawyer employed by the Canadian Association of University Teachers (CAUT) for an opinion regarding my case. The CAUT lawyer replied, strongly recommending that the faculty association mount a grievance on my behalf. Her letter concluded with the following advice:

> I would recommend that the Association agree to grieve this retirement on the basis of discrimination in employment on the prohibited ground of "sex," not on the ground of "age."
>
> The argument would be that the "normal" age of retirement is discriminatory on the ground of sex insomuch as it aggravates the effects of systemic discrimination against women in employment....

> Additionally, it is discriminatory more specifically in this case as (you will need evidence on this) most male faculty have longer careers than the griever (and other women at King's and/or UWO) in this case, who, as a woman, commenced employment late in life. Consequently, "normal" is itself discriminatory as it "advantages" men predominantly who generally have more years in service prior to retirement, resulting in both longer period of employment income and a healthier retirement income at the end of their career....

When members of the faculty association met to discuss my case on September 24, they had received two legal opinions, both arguing in favour of a grievance, albeit on different grounds. Janice Rubin recommended that they proceed on the basis of age discrimination, and the CAUT lawyer cited strong evidence of gender discrimination in my case. After lengthy debate, however, the majority of members voted against pursuing a grievance on my behalf. On Wednesday, October 8, my lawyer received a letter from the president, providing a brief description of the association's views and the rationale behind its decision:

> Firstly, although the Human Rights Code may not be in keeping with the Charter, King's College does not have a duty to go beyond the Code nor to question the Code's constitutionality. Secondly, although there is no explicit mandatory retirement clause in our Terms and Conditions, there is considerable evidence of an explicit understanding between the Faculty Association and the College as to the practice of mandatory retirement as a working assumption in many of our other agreements.

My lawyer sent a letter to the faculty association executive expressing regret that members had voted against a grievance procedure. She then began the process of launching a formal complaint with the Ontario Human Rights Commission, naming King's College as respondent. The complaint was lodged in late November and the college lawyer returned a Response to Complaint to the Commission, requesting the complaint be dismissed, "in as much as the complaint is frivolous and vexatious and filled [*sic*] without reasonable grounds to believe that discrimination has occurred.... "

Going Public

In December 2003, after consultation with my lawyer, I decided to tell my story to the public. Paul Martin had just been sworn in as prime minister and had announced that he favoured an end to mandatory retirement, media interest in the issue had increased, and rumours had begun to circulate about the new Ontario Liberal government's intention to end the practice. Almost all debate centred on human rights in general, ignoring the negative impact of mandatory retirement on women. It was time to add my voice to the public discussion.

Janice Rubin and I were interviewed in her office for an article published in the January 30 *Globe and Mail* "Report on Business" section.[20] This presented a very fair account of my case, indicating that I wanted to be reinstated to my faculty position at King's College. It also reported my lawyer's intention to proceed with the challenge if the Ontario government abolished mandatory retirement without consideration for my case and others like it, emphasizing the especial injustice of mandatory retirement for women whose pensions were reduced because of interrupted work histories or delayed entry into the labour force. Since then, a number of articles in major Canadian newspapers have repeatedly asserted that mandatory retirement is unfair.[21]

My complaint to the Ontario Human Rights Commission was slowly working its way through the system. On March 8, 2004, Janice Rubin received a copy of the case analysis report prepared by the Ontario Human Rights Commission Mediation Office in response to the request of the respondent (King's College) to reject my complaint. The college's lawyer cited subsections 34(1)(a), 34(1)(b) and 34(1)(c) of the Human Rights Code as grounds for rejecting my request for reinstatement. Subsection 34(1)(a) states that "the complaint is one that could or should be more appropriately dealt with under an Act other than this Act," and the mediation team agreed that the provisions of the Ontario Labour Relations Act, 1995, could more appropriately deal with my case—but concluded that this route was not possible because my bargaining unit (the King's College Faculty Association) had voted against pursuing a grievance on my behalf. Subsection 34(1)(b) provides that "the subject matter of the complaint is trivial, frivolous, vexatious or made in bad faith." The team rejected this argument from the respondent, citing the *Shorter Oxford English Dictionary*'s definition of "frivolous" as "of little or no weight or importance … characterized by lack of seriousness" and arguing that, by this definition, the complaint could not be characterized as "frivolous." Subsection 34(1)(c) allows that "[if] the complaint is not within the jurisdiction of the Commission … the Commission may, in its discretion, decide not to deal with the complaint." Consequently, the mediation team's final recommendation suggested "the Commission exercise its discretion and decide not to deal with this complaint pursuant only to subsection 34(1)(c) of the Code." In effect, the Commission's hands are tied because it does not deal with people aged sixty-five or older. Apparently we do not enjoy the same basic human rights as younger people.

One of the frustrations of this case lies in the slow progress of documents through the hands of Ontario Human Rights officials, especially as the end of my first year of "retirement" approached with no immediate sign of resolution of my case. The year had been instructive in many ways because I found that people tended to subscribe to specific images of appropriate retirement behaviour. Many suggested I might feel some regret about mounting the challenge—I must have found retirement to be surprisingly pleasant and surely would not want to return to my former job. Indeed, I enjoyed the year very much because of the freedom to set my own schedule and to engage in some activities that would have been impossible during a busy teaching term. But the year had been like a welcome sabbatical, providing an opportunity to write without interruption. Far from feeling like a retiree, I had continued to spend most mornings in my home office, working on academic papers and a forthcoming book.[22] I presented papers at two conferences during 2004, so it was a very productive time. I continue to expand my research horizons. Simply put, this is what I "do." This aspect of academic life, more than anything else, sets us apart from many other occupations. We deal in knowledge, and ideas do not stop flowing when retirement is imposed.

Media interest in the mandatory retirement debate is ongoing. Immediately after the *Globe and Mail* article appeared in January, my lawyer received a call from the Canadian Broadcasting Corporation. On March 31, I was interviewed by CBC Radio One's morning show, *Sounds Like Canada*. Like the newspaper article, the program was well-balanced and fair. One of the overriding concerns with respect to universities is the fear that elimination of mandatory retirement would allow incompetent professors to continue working—or, in the words of most critics, "How would we get rid of them?" The interviewer posed this question along with a query about the evaluation of older professors. In my opinion, the responsibility for dealing with both these potential problems lies with the senior administrative officers within our respective institutions. Our credibility in the classroom and in the scholarly literature is confirmed by student and peer evaluations. If we fall short of expected performance standards at any age, the university has an obligation to deal with these shortcomings, not to wait until age sixty-five to "get rid of" a faculty member with a history of problems, and in the process punish other, more effective teachers and scholars by retiring them.

CBC Television interviewed me for a story on *The National*, which aired on May 11, 2004. Once again, the media presentation appeared to favour the elimination of mandatory retirement; in the allotted three-minute segment, the commentary emphasized the fact that many sixty-five-year-olds defy the stereotypical view that we are physically and mentally challenged. Comments from younger colleagues were not included, but the principal of the college appeared briefly, arguing that older faculty members with high salaries should retire and clear the way for young people ready to make their own unique contributions to academic life. My interview gave me a chance to emphasize the unfairness of a practice that discriminates against willing workers solely on grounds of age.

Conclusion

On May 10, 2004 the Ontario Human Rights Commission sent a letter to Janice Rubin, citing subsection 34(1)(c) of the Human Rights

Code as the reason they could not deal with my complaint. Until my case is complete, it will not be possible to draw definitive conclusions about the elimination of mandatory retirement, or about the part my challenge will have played as one link in a much longer chain of events.

My best summation is to acknowledge and celebrate a social and political climate that appears ready to honour the basic human rights of all people, regardless of age. My case is unique to the extent that my career has been truncated, leaving me with a relatively small pension, but many other women (and some men), lacking desire or means to protest, have found themselves in similar positions. Why, then, have I chosen to accept an additional financial burden by engaging legal counsel to mount the challenge when, given the slow pace of the judicial process, I may never reap any personal benefit? I chose this route because I found the whole matter so abundantly unfair that I felt constrained to do something, and there has been a certain satisfaction in taking action in response to the powerlessness that I felt within an institution bound by archaic rules about age and performance. If my action helps to further the cause against mandatory retirement, then I will feel to some extent vindicated, secure in the knowledge that my complaint is neither "frivolous" nor "vexatious." It is a very serious pursuit taken after much thought about the costs and the consequences.

12

Conclusion

Peter H. Russell

A New Social Movement

At the October 2002 founding conference of AROHE (Association of Retiree Organizations in Higher Education) in Bloomington, Indiana, I was surprised to hear several speakers refer to our associations as part of a social movement. As a political scientist I was familiar with the concept of "gray power," but the social movement referred to went well beyond participation in democratic politics. It was, and is, a movement akin to the movement for women's liberation, the major social revolution in industrialized countries in the twentieth century. This movement is aimed at changing the status of a large section of the population, workers who have reached what is conventionally regarded as the age of retirement—freeing them from unfair and impractical restrictions on their opportunities. The organization and publication of this book indicate this new social movement is coming into its own in Canada. As with any respectable academic project, the book pays attention to both sides of arguments and recognizes the obstacles to be overcome and the limited value of any progress that is made. But the direction of reform reflected in this book is clear, particularly on what has emerged as the movement's cutting-edge issue—the banning of mandatory retirement. In law, public opinion, and politics, momentum is building against policies that force Canadians, regardless of their abilities or desires or society's needs, to give up their jobs solely because they have celebrated their sixty-fifth birthdays.

Like feminism, the movement against ageism challenges forms of discrimination systemic to the organizing structures of society. Unlike feminism, it is propelled as much, if not more, by demography than by ideology. It would be more accurate to say that it is propelled by the demographic consequences of modern medicine. Without the huge increases in life expectancy that medical science and relative affluence have made possible in the industrialized west, there would be no people to participate in a movement against ageism. The chapter by Don Kerr and Roderic Beaujot presents the Canadian data on this phenomenon. Whereas in 1986, 10 percent of Canada's population was over sixty-five, it is projected that by 2036 a quarter of the population will be over sixty-five. Kerr and Beaujot observe that failure to reform employment and social policy will mean the fastest-growing part of Canada's population will have a drastically low participation rate in its labour force. Consigning over-sixty-fives to society's scrap heap, altogether aside from its negation of fundamental human rights, is a policy the country can no longer afford.

The Convergence of Social Justice and Economic Utility

Indeed, it is the convergence of human rights and economic considerations that now provides the basis for a breakthrough and reform. Terry Gillin reminds us in Chapter 1 that in the past the struggle against mandatory retirement pitted human rights arguments against arguments for economic efficiency. The remnants of those economic arguments can be found in the chapter by Morley Gunderson and Douglas Hyatt (Chapter 8). Other economics contributors, notably Jonathan Kesselman and John Munro, cast doubt on the cogency of the economic case for mandatory retirement as put forward three decades ago, and point out how labour market and demographic changes undermine some of its key assumptions. Acute shortages of qualified persons for jobs in many sectors of the economy, including universities, render the classic argument that forced retirement at age sixty-five is the only way to open up jobs for the young, in Kesselman's words, "particularly archaic." Munro exposes the bias and lack of evidence in assumptions about the declining productivity of older workers, particularly in fields where experience and intellectual acumen are desirable qualifications. If the traditional economic arguments against voluntary retirement were valid, one would expect to have seen by now dire economic consequences in jurisdictions (like the United States and Quebec) that abolished mandatory retirement some time ago. As a number of contributors point out, there is no sign of these predicted adverse effects in any of these places. I should add that it surely would be surprising to find the United States sacrificing economic competitiveness to a labour policy more compatible with human rights.

While demographics and economics have produced a much more favourable political environment for reforming labour and other policies that regulate the right to work, it is the appeal to fundamental human rights that provides the moral energy for the sustained attack on systemic discrimination against older citizens. This is always the case with movements that effect major changes in social structure and economic policy. Democratic leaders will only support change when convinced it is in the broad public interest to do so. They will not sacrifice the economic interests of the majority to honour and safeguard the rights of a minority. That is why the changing economic and demographic circumstances are essential conditions for policy reform. It is those who directly experience the hurtful and unfair deprivation of their rights, and others sensitive to their plight, who provide the political momentum, pressing politicians and lawmakers to commit to reform and keeping them from backing off of their commitment or unduly delaying its fulfillment.

Anyone with any doubt about how hurtful ageist policies can be should read Jean McKenzie Leiper's contribution to this book (Chapter 11). Her story illuminates the human side of a state of affairs too often discussed in impersonal statistical terms. There would surely have to be some general public interest of the most compelling nature—which there is not—to justify terminating this woman's employment. It is probably true, as a number of contributors point out, that only a minority of those affected by mandatory retirement, perhaps between 10 and 20 percent, are "involuntarily" retired. But of course that fact does not detract from the cogency of Dr. McKenzie Leiper's human rights claim. The rights civilized societies treat as fundamental are very often of vital concern to only a minority of their members. Comfortable majorities do not need rights protection.

In Chapter 8, Gunderson and Hyatt raise the issue of the collective right of unions and other organized groups of employees to enter, consensually, into contracts with employers to establish a mandatory retirement age. It was probably largely through such agreements that mandatory retirement was introduced in Canada—although, as Munro reports, this was certainly not the case at his own university, where mandatory retirement was first established by administrative fiat. But assuming that in many places of employment mandatory retirement was instituted through a collective agreement and might be maintained in that way in the future, how should we assess the collective right of the majority to curtail the freedom of a minority of their colleagues? I believe a strong commitment to collective bargaining rights has much to do with organized labour's ambivalent attitude to the movement to abolish mandatory retirement, as reported in Chapter 4 by Thomas Klassen with David Forgione. Labour's ambivalence and, in some instances, open hostility to ending mandatory retirement partly accounts for the fact that, thus far, only one jurisdiction in Canada, Quebec, has enacted an absolute prohibition of contractual mandatory retirement. So the question arises: should the freedom of the majority to negotiate their terms of employment be permitted to trump the freedom of individuals to decide when to terminate their employment?

In answering this question, we must distinguish between two very different explanations of why a majority might favour retention of mandatory retirement. One reason is that the majority is made up of those happy about the prospect of retirement at the stipulated age plus others so far from retirement they give the matter little thought and are willing to go along with the first group. In this situation, a combination of majority complacency and indifference should not be permitted to force the minority to give up their freedom. The more serious reason for a majority to favour mandatory retirement is the belief that mandatory retirement is tied, inexorably, to their pension rights. Klassen and Forgione report that concern about retaining defined benefit pensions is high among that part of organized labour which favours retention of mandatory retirement. The answer to this concern, well articulated by Munro in the concluding pages of Chapter 10, is that legislation prohibiting contractual mandatory retirement must protect employee rights to retire with full pension and other benefits at the age of retirement stipulated in existing contracts of employment. If that condition is met—and it may well be the most difficult condition to satisfy in enacting reform legislation—then the case for legally condoning contractual mandatory retirement collapses.

There is the further possibility that though employers may be willing or legally bound to retain existing pension rights, they may anticipate higher pension costs and demand major concessions on other remuneration issues. The answer, at least in part, is to insist on a realistic appraisal of the effect of ending mandatory retirement upon the average age of retirement. David MacGregor (Chapter 2) presents evidence that where mandatory retirement has been abolished, the age at which people retire has changed very little. Even if incremental pension costs are small, some employers, and none more than university administrations faced with chronic public underfunding, will likely push this point. To guard against the younger majority being pressured to pay a significant price for their elders' freedom to continue employment after age sixty-five provincial governments should be asked to fund additional costs attributable to the abolition of compulsory retirement. Governments may well resist such a reasonable request—they have been known to do so before. In this case, the younger members of the employee group should reflect on the reasonableness both for themselves and their older colleagues of paying a price for gaining freedom and flexibility in making their long-term career plans.

Legislative, Judicial, and Negotiated Remedies

Jean McKenzie Leiper decided to initiate a legal challenge to the action of her employer, the University of Western Ontario's King's College, in terminating her employment solely because she reached age sixty-five. She acknowledges that a court case will take a very long time and cost her a lot of money, with no guarantee of success. But a more expeditious way of securing justice was foreclosed when a majority in her faculty association refused to support her use of the local grievance procedure to challenge the college's retirement policy. Her story demonstrates the unwillingness of employee and administration leaders to exercise leadership in instituting reforms long overdue. Dr. McKenzie Leiper's colleagues, and her university's officers, prefer to wait for the provincial legislature to render existing mandatory retirement policy illegal. The current Ontario government has declared itself "committed to giving Ontario workers the right to choose when they retire," and to doing so "without undermining existing retirement rights or entitlements to benefit and pension plans."[1] Even if the government of Ontario honours this commitment reasonably soon, McKenzie Leiper's court action, and others like it, may still be necessary for those who are too late to benefit from the legislative change or who live in other, less enlightened, jurisdictions.

If Dr. McKenzie Leiper's case does reach the Supreme Court of Canada, it stands a pretty good chance of success. Chapter 3, by Terry Gillin and Thomas Klassen, gives a full account of the roller coaster ride of challenges to ageism in the Supreme Court of Canada. As a commentator on the Supreme Court's performance, I have used the roller coaster image before—in the context of its treatment of Aboriginal rights.[2] Though these ups and downs and swings back and forth in judicial decision-making may shock legal purists, they do not surprise social science observers of the treatment of contentious policy issues by high constitutional courts. In the long run, in virtually all constitutional democracies, judicial decision-making on these issues follows election returns. So there is hope for Dr. McKenzie Leiper and others like her now turning to the courts for relief from age-based discrimination.

Judicial adjustments to the spirit of the times are all the more to be expected in applying Canada's Charter of Rights and Freedoms, which requires the judiciary to conduct a two-stage inquiry. First, judges must ascertain whether the challenged legislation or government action infringes a Charter right or freedom. If the answer is in the affirmative, the Charter, recognizing that no right can be treated as absolute and without limits, requires the court to ascertain whether the limit on the right is "reasonable" and "demonstrably justified in a free and democratic society." If the answer to the second part of the inquiry is in the affirmative, the court will uphold the law or action even though it encroaches upon a fundamental right. In performing the Charter two-step on challenges to mandatory retirement and other laws and practices that deny Canadians rights or benefits solely because of age, the Supreme Court has been consistent and unanimous in answering the first question: Such legislation or policy violates the right enshrined in section 15 of the Charter of "every individual to … equal protection and equal benefit of the law without discrimination and, in particular, without discrimination based on … age…."[3] Thus we can take it as established constitutional law that laws condoning mandatory retirement and other age-based discrimination against older workers encroach on Canadians' constitutional rights. It is in the second part of the inquiry, where they consider policy-based arguments for limiting older workers' rights, that a majority of Supreme Court justices have supported arguments in favour of mandatory retirement. Mr. Justice Gérard La Forest, writing for the majority in the 1990 *McKinney* case, recognized what he took to be the spirit of his time and pronounced in dicta cited above by Gillin and Klassen that "65 has become generally accepted as the 'normal' age of retirement" and that "mandatory retirement has become part of the very fabric of the organization of the labour market in this country."

It is unlikely that a majority of Supreme Court justices or lower court judges, in 2004 or beyond, facing a challenge to the same laws and policies that were before Justice La Forest, will view the spirit of the times in the same way. Court decisions since the Supreme Court's rejection of Olive Dickason's challenge to mandatory retirement in 1992 have been showing greater sensitivity to the human rights aspects of policies restricting older workers' employment opportunities and tougher scrutiny of justifications for age-based dismissals, including rationales based on bona fide occupational requirements. Gillin and Klassen also show that legislative changes across the country in recent years have been in the direction of treating mandatory retirement as a legally prohibited form of discrimination. They point out further that Canada's sister constitutional democracies, major trading partners, and economic competitors are abandoning mandatory retirement. Other chapters present evidence and analyses in an effective rebuttal of arguments in favour of mandatory retirement that Justice La Forest and four colleagues found convincing fifteen years ago. The case is particularly convincing with respect to arguments about the need to provide jobs for younger workers and broad generalizations about the declining capabilities of older people.

The one argument in support of compulsory retirement that continues to have some appeal is more in the nature of human rights—the contention that if retirement is voluntary, employers will introduce harsh and humiliating competency tests into the workplace to weed out unproductive employees. In the context of university employment, where most faculty are tenured, mandatory retirement at a fixed age, it is argued, is the only kind way of getting rid of the incompetent. This argument, as John Munro observes, is based on a mistaken view of tenure policy, which, if properly administered, does not guarantee continued employment to the incompetent and dissolute. Munro also points to the performance tests built into personnel practices at universities and other workplaces. To this I would add the personal observation that since the mid-1970s universities have been hiring their faculty in a highly competitive and merit-oriented manner—a far cry from the "old boys' network" that prevailed when I was hired in the 1950s. Many of these colleagues are at the very top of their professional careers just as the age-sixty-five guillotine threatens to lop off their heads. No wonder more and more of them are leaving to take up positions at universities in jurisdictions that are not addicted, as David MacGregor puts it, to such a lazy way of planning. Granted that even though experience has shown that where retirement is voluntary the great majority retire before age seventy, a few old fossils will hang on and something will have to be done with them—and whatever is done may be painful. But against this occasional adverse consequence of a regime of voluntary retirement we must weigh the certain and extensive adverse consequences of a regime of mandatory retirement. Passages of Madam Justice L'Heureux-Dubé in her dissenting decisions in *McKinney* and *Dickason*, cited by Munro, are most eloquent:

> Mandatory retirement arbitrarily removes an individual from his or her active worklife, and source of income, regardless of his or her actual mental or physical capacity, financial wherewithal, years of employment in the work force, or individual preferences.... Forced retirement from the work force strictly on account of age can be extraordinarily debilitating for those entering their senior years.

Bearing in mind that the Supreme Court in its leading case on justifiable limits of Charter rights has said that the onus is on government to demonstrate to "a very high degree of probability" that the limit is justified by a "pressing and substantial" objective and that the means used are proportionate to that objective,[4] if Dr. McKenzie Leiper or others in her position ever come before the Supreme Court their prospects for success are very high. That said, it will be a pity if Canadians who suffer the adverse consequences of forced retirement have to wait the two or three years it will take to bring their cases before the Supreme Court of Canada. The injustice and personal harm these people are experiencing is real and unnecessary. The best remedy, the one immediately at hand and most consensual, is for employers and employee organizations, rather than waiting to be forced by the legislature or the judiciary to do the right thing, to negotiate a new set of retirement policies that serve both parties better than the existing antiquated regime.

The Larger Picture

At present, ending forced retirement is the front line issue in a much larger movement. This movement is nothing less than a challenge to rethink and restructure the way we organize what is becoming for many the final third of their lives. It clearly won't do to continue to treat this third period as one of idleness, disengagement, and dependence.

The contribution of Norene Pupo and Ann Duffy to this book

opens up this larger picture. Indeed, as they argue, grappling with the rigidities and mindset that now afflict the lives of seniors can be "a catalyst for rethinking the complex inter-connections not only between work and family, but between work, community service, volunteerism, and leisure."

When retirement for those in our society who are fortunate enough to have secure long-term jobs becomes voluntary, these people will still "retire." Evidence from jurisdictions that have made this transition indicates the average age of retirement increases only marginally after abolition of mandatory retirement and may actually decline, particularly if attractive post-retirement options are in place. There is also evidence that a large number of persons who "retire" do not look forward to a total termination of "work." One of the principal objectives of retiree organizations developing in the sphere of post-secondary education is to establish programs and structures that will enable retirees to continue contributing to the academic and professional activities of universities and colleges, as well as to the larger community. Academic administrators are just now waking up to the possibility that the approximately 40 percent of their faculty who will retire in the next decade need not be "lost" to the institution. At the founding meeting of the University of Toronto's retiree association, a colleague expressing skepticism about what such an organization could do said that since his retirement the university had treated him like a "non-person." Far too many retirees have had similar traumatic experiences.

Pupo and Duffy report on a case study of eighteen retirees who re-entered the work force after retirement. Besides indicating the creativity and productivity of older workers, this study shows how disinclined older Canadians are to sit around and do nothing for the final third of their lives. "Freedom 55" may be an effective advertising slogan, but it is an extremely inaccurate portrayal of the aspirations of those who reach age 55—or 65. Pupo and Duffy acknowledge that the retirees in their case study are not a cross-section of the population. They were all in receipt of pension income, had higher than average levels of education and affluence, and were undertaking "post-retirement" work that varied from self-employment to part-time and full-time employment for short periods of time. Less fortunate older workers—and these are probably the majority—face a different set of problems. Given the bias against hiring older workers, too many face a labour market in which the only employment available consists of non-standard, part-time, low-paying jobs.

A key to reform is overcoming the rigidities that structure the organization of work and jobs. It does not make sense to have as the standard of employment big full-time jobs that burn people out, leave little time for family or leisure or community volunteering, continue for many years, then suddenly, at retirement, come to a full stop. Chapter 7, by Norene Pupo and Ann Duffy, points to one way of introducing more flexibility into the world of work: the adoption of phased retirement programs enabling workers to "retire" early but continue to work on a part-time basis while collecting pension benefits. This kind of restructuring can open up positions for younger workers while retaining the experience and skills of older workers. Of course, its attractiveness, as Pupo and Duffy make clear, depends on securing equal rates of pay, security, and benefits for full- and part-time employees.

As we venture down this road and begin to move away from structuring our lives into sharply divided periods of work and retirement, nothing less than a new vocabulary will be needed. As we recognize that "work," altogether aside from generating income, gives people a sense of personal fulfillment and provides the setting for much of their social life and engagement, and that "retirement" cannot mean thirty years of leisured disengagement from these aspects of work, we will need to develop new ways of thinking about these phases of our lives. But we will not be in a position to take up these challenges in an effective and equitable manner until the biased stereotyping of older citizens is overcome. And the essential step towards accomplishing this is removing from Canadian statute books laws that condone discrimination against workers who have reached age sixty-five. That is why the abolition of laws permitting mandatory retirement is so important and so urgent. Let us hope this book will serve as an inspiration to legislators to take that essential step soon, so we can get on with the more creative and challenging reforms that lie ahead.

Notes on Contributors

Roderic Beaujot is Professor of Sociology at the University of Western Ontario. He received his PhD from the University of Alberta (1975), and spent 1974–76 at the Demography Division of Statistics Canada before going to Western in 1976. His *Earning and Caring in Canadian Families* (Broadview, 2000) won the Porter Prize from the Canadian Sociology and Anthropology Association. He is the co-author, with Don Kerr, of *Population Change in Canada* (Oxford, 2004).

Ann Duffy is a Professor of Sociology and Labour Studies at Brock University in St. Catharines, Ontario. She has worked extensively in the area of non-standard employment, focusing in particular on women and part-time work. Professor Duffy, with colleague Professor Norene Pupo, is currently researching the implications of part-time employment for seniors in Canada.

David Forgione recently completed his MA in the Department of Political Science at York University.

C. T. (Terry) Gillin is a professor in the Department of Sociology at Ryerson University.

Morley Gunderson is the CIBC Professor of Youth Employment at the University of Toronto and a professor at the Centre for Industrial Relations and the Department of Economics. He is also a research associate of the Institute for Policy Analysis, the Centre for International Studies, and the Institute for Human Development, Life Course and Aging, and an adjunct scientist at the Institute for Work and Health.

Douglas Hyatt is a professor at the Joseph L. Rotman School of Management, the Centre for Industrial Relations, and the Institute for Policy Analysis, all of the University of Toronto.

Allan Irving is a professor in the School of Social Work at King's University College at the University of Western Ontario. Among other publications, he is the author of *Brock Chisholm: Doctor to the World* (1998), and co-editor of *Reading Foucault for Social Work* (1999). He is currently completing, with a colleague, a book on social work practice in the postmodern era to be published by Columbia University Press.

Don Kerr is an associate professor of Sociology at King's University College, which is affiliated with the University of Western Ontario. He was previously a population analyst at the Demography Division at Statistics Canada.

Jonathan R. Kesselman is a professor in the Public Policy Program, Simon Fraser University, Vancouver, and holds the Canada Research Chair in Public Finance. From 1972 through 2003, he was a professor in the Department of Economics, University of British Columbia.

Thomas R. Klassen is assistant professor in the Department of Political Science at York University, and Coordinator of the Public Policy and Administration Program. He is the author of *Precarious Values: Organizations, Politics and Labour Market Policy in Ontario* (McGill-Queen's University Press, 2000).

Jean McKenzie Leiper is completing a book on time in the lives of Ontario women lawyers. Until July 1, 2003, she was an associate professor in the Department of Sociology at King's University College at the University of Western Ontario.

David MacGregor is a professor of sociology and Chair of the Department of Sociology at King's University College at the University of Western Ontario. He is the author of *Hegel and Marx After the Fall of Communism* (1998) and *Hegel, Marx and the English State* (1992). His *The Communist Ideal in Hegel and Marx* (1984) won the 1986 Porter Prize of the Canadian Sociology and Anthropology Association.

John Munro is Professor Emeritus of Economics at the University of Toronto, having been subjected to mandatory retirement June 30, 2003, after thirty-nine years of teaching and publishing in European economic history (1250–1914). He recently served as one of the area editors of the five-volume *Oxford Encyclopedia of Economic History* (New York, 2003); and continues to serve on the Executive Board of the *Istituto Internazionale di Storia Economica "Francesco Datini" da Prato*, in Italy.

Norene Pupo is Director of the Centre for Research on Work and Society and an associate professor of Sociology at York University. She is principal investigator of a SSHRC-funded research alliance, Restructuring Work and Labour in the New Economy. Along with colleague Professor Ann Duffy, she is studying the impact of globalization on work-time arrangements and in particular is considering shifts in work patterns amongst women in service industries.

Peter H. Russell taught political science at the University of Toronto for thirty-eight years until his retirement in 1996. He has published widely in constitutional, judicial, and Aboriginal politics. He was the founding President of RALUT (Retired Academics and Librarians of the University of Toronto), and is the founding president of CURAC (College and University Retiree Associations of Canada).

Notes

Chapter 1 Introduction

1 See Special Senate Committee on Retirement Age Policies. David A. Croll, Chairman. 1979. *Retirement Without Tears*. Hull, (Quebec: Minister of Supply and Services Canada, 1979).

2 See J.E. Pesando, *The Elimination of Mandatory Retirement: An Economic Perspective* (Toronto: Ontario Economic Council, 1979).

3 Another influential early voice was the American economist Lazear, who argued that employers tend to underpay workers, relative to their productivity, early in their careers and overpay them during their later years, suggesting a justification for mandatory retirement. See Edward Lazear, "Why is There Mandatory Retirement?" *Journal of Political Economy* 87:6 (1979), 1261–84.

4 By virtue of including university professors in its law, Quebec has the most comprehensive provincial legislation banning mandatory retirement. It is perhaps for this reason that the scholarly debate has taken place in English Canada.

5 Ontario Task Force on Mandatory Retirement (R. Ianni, Chairman), *Fairness and Flexibility in Retiring from Work: Report of the Ontario Task Force on Mandatory Retirement* (Toronto: Ontario Ministry of Labour,

1987), 28. The task force did not arrive at a consensus on whether a legislative ban on mandatory retirement was required. The chair of the commission supported such a formal ban, while the two other members of the task force supported greater flexibility in retirement but no formal government action at that time (29).

6 See Gillin and Klassen, Chapter 3.

7 *McKinney v. University of Guelph*, [1990] 3 S.C.R. 229.

8 See Morley Gunderson and James Pesando, "Eliminating Mandatory Retirement: Economics and Human Rights," *Canadian Public Policy* 6:2 (1980), 352–60.

9 See Bibliography.

10 Using 1979 (the publication of *Retirement Without Tears*), 1990 (the *McKinney* decision) and 2000 (more arbitrarily) as benchmarks, we estimate that between 1979 and 1989, thirteen scholarly reports and publications on mandatory retirement in English-speaking Canada were published. Between 1990 and 1999, there were ten such publications, and since 2000, thirteen more have been published. Gunderson, either as author or coauthor, has published at least ten articles of the overall total of thirty-six, four since 2002.

11 Canadian Human Rights Review Panel, *Promoting Equity: A New Vision* (Ottawa: Canadian Human Rights Review Panel under the authority of the Minister of Justice and the Attorney General of Canada, June 23, 2000).

12 The panel was headed by then-retired Justice La Forest, who wrote the majority opinion in the 1990 *McKinney* case.

13 Ontario Human Rights Commission, *Time for Action: Advancing Human Rights for Older Ontarians* (Toronto: Ontario Human Rights Commission, 2001); see also Arlene T. McLaren and Margaret M. Mancry, *Factors Affecting the Economic Status of Older Women in Canada: Implications for Mandatory Retirement* (Victoria: British Columbia Human Rights Commission, 2001).

14 See Gillin and Klassen (Chapter 3) for discussion of *Greater Vancouver Regional District Employees' Union v. Greater Vancouver Regional District*, [2001] BCCA 435.

15 Richard Brennan, "Tories May Scrap Forced Retirement," *Toronto Star*, 6 April 2003. The core of the proposed legislation (Bill 68,

Mandatory Retirement Elimination Act, 2003 [37th Legislature, 4th Session]) was to redefine age in the Ontario Human Rights Code. Currently, for the purposes of discrimination in employment, age is defined as more than 18 and less than 65 years. This has made it possible for employers and unions to require employees to retire at age 65. Bill 68 would have removed the ceiling of 65 years.

16 Canadian Press, "McGuinty Promises Bill to Ban Forced Retirement," *Toronto Star*, 29 January 2004.

17 Susan Delacourt, "Martin Opposes Mandatory Retirement," *Toronto Star*, 20 December 2003, A10.

18 Buzz Hargrove, "Mandatory Retirement Debate," *The Globe and Mail*, 29 December 2003, Comment.

19 Heather Scoffield, "Rethink CPP's Age-65 Rule, Central Bank Chief Urges," *The Globe and Mail*, 21 April 2004.

20 Ministry of Labour, *Providing Choice: A Consultation Paper On Ending Mandatory Retirement* (Toronto: Ministry of Labour, August 2004).

21 Robert N. Butler, "Ageism: Another Form of Bigotry," *The Gerontologist* 9:3 (1969), 243–6. Butler coined the term "ageism." For a current perspective, see Todd D. Nelson, ed., *Ageism: Stereotyping and Prejudice Against Older Persons* (Cambridge, MA: MIT Press, 2003).

22 Donald G. McTavish, "Perceptions of Old People," in *Research Instruments in Social Gerontology: Clinical and Social Psychology* (vol 1), ed. David J. Mangen and Warren A. Peterson (Minneapolis: University of Minnesota Press, 1982). McTavish reviews a number of studies of attitudes toward being old.

23 Research generally refutes the prevailing myths about older workers; see Donald H. Kausler, "Motivation, Human Aging, and Cognitive Performance," in James E. Birren and K. Warner Schaie, eds., *Handbook of the Psychology of Aging*, 3rd ed., (San Diego, CA: Academic Press, 1990), 172–83; Pauline K. Robinson, Sally Coberly, and Carolyn E. Paul, "Work and Retirement," in Robert H. Binstock and Ethel Shanas, eds., *Handbook of Aging and the Social Sciences*, 2nd ed. (New York: Van Nostrand Reinold Company, 1985), 503–27; and Waneen W. Spirduso and Priscilla Gilliam MacRae, "Motor Performance and Aging," in James E. Birren and K. Warner Schai, eds., *Handbook of the Psychology of Aging*, 3rd ed. (San Diego, CA: Academic Press, 1990), 184–200. In a

recent review of the literature, Meadows argues that, except for a limited number of jobs, work performance does not deteriorate with age—at least up to age 70, after which few are employed and scientific evidence is not available. See Pamela Meadows, *Retirement Ages in the UK: A Review of the Literature*. Employment Relations Research Series No. 18 (Department of Trade and Industry, 2003) Crown copyright. *http://www.dti.gov.uk/er/emar/errs18.pdf.* (2 August 2004).

24 C. Wright Mills, in *The Sociological Imagination* (New York: Grove Press, Inc., 1959), wrote about personal troubles and public issues.

25 According to David Foot, "Demographics explain about two-thirds of everything," David K. Foot with Daniel Stoffman, *Boom, Bust & Echo: Profiting from the Demographic Shift in the New Millennium* (Toronto: Macfarlane Walter & Ross, 1998), 8.

26 It needs to be reiterated that Canada does not have a unified mandatory retirement policy. As summarized by Gillin and Klassen in Chapter 3, there are a variety of rules across the provinces, territories and federal government.

Chapter 2

1 Aesop, *Aesop's Fables*, based on the translation of George Fyler Townsend, with an introduction by Isaac Bashevis Singer. Illustrated by Murray Tinkelman. New York: Doubleday, 1968.

2 Jeff McMaster, "Mandatory Retirement: Is It Finally Time to Send the Policy off to Pasture?" *Folio* (Alberta: Public Affairs Department, University of Alberta, 2004), 5.

3 Sandra Acker, "Women Academics and Faculty Shortages: Through the Lens of Gender Analysis," *Ontario Confederation of Faculty Associations Forum* (Toronto: OCUFA, 2004). These figures are from the Association of Universities and Colleges of Canada (AUCC), *Trends in Higher Education* (Ottawa: AUCC, 2002).

4 Simone de Beauvoir, *The Coming of Age* (New York: G.P. Putnam's Sons, 1972), 263.

5 The University of British Columbia Senate Ad Hoc Committee on the Academic Implications of Mandatory Retirement at Age 65, *Re: Information on Mandatory Retirement for the May Meeting of Senate*, 3 May 2002.

[6] *Bell Canada v. Office and Professional Employees' International Union, Local 13*, May 28, 1973. Quoted in Daniel Jay Baum, *The Final Plateau: The Betrayal of Our Older Citizens* (Toronto: Burns and MacEachern Limited, 1974), 155.

[7] Grant Schellenberg and Cynthia Silver, "You can't always get what you want: Retirement preferences and experiences," *Canadian Social Trends*, No. 75, Winter 2004, Statistics Canada, Cat-No. 11-008.

[8] Jeffrey Cormier, *The Canadianization Movement: Emergence, Survival, and Success (*Toronto: University of Toronto Press), 2004.

[9] Colin Freeze and Marina Jimenez, "Strippers put Ottawa program at centre stage," *The Globe and Mail*, 27 November 2004.

[10] Policy Research Initiative, Population Aging and Life-Course Flexibility: The Pivotal Role of Increased Choice in the Retirement Decision, Discussion Paper (Ottawa: Government of Canada, 2004), 18.

[11] Statistics Canada. *General Social Survey: Social Support and Aging, 2002*: Cat. No. 89F0115XIE, Ottawa, September 2, 2003.

[12] Government of Canada: Policy Research Initiative, *Population Aging*, 18.

[13] Doreen Duchesne, "More Seniors at Work," *Perspectives on Labour and Income*, Statistics Canada, Cat. No. 75-001XIE 5:4 (February 2004), 7.

[14] Ken MacQueen, "The Dawn of a New Old Age," *Maclean's*, 24 May 2004.

[15] Peter Brieger, "Extended Retirement Age Welcomed by Rights Advocates," *Financial Post*, 2 May 2003.

[16] Richard Brennan, "Tories Set to Unveil Retirement Rules. Mandatory Departure at 65 to Be Eliminated, sources say," *Toronto Star*, 17 April 2003.

[17] Buzz Hargrove, "Keep Mandatory Retirement," *The Globe and Mail*, 29 December 2003; Brieger, "Extended Retirement Age Welcomed;" "Critics Blast Retirement Ban: Tory Proposal to End Mandatory Age Requirement Could Harm Youth Employment," *Windsor Star*, 2 June 2003.

[18] Schellenberg and Silver, "You can't always get what you want," 6–7.

[19] *GVRDEU v. Greater Vancouver Regional District* (2001), 206 D.L.R. (4th) 220.

[20] Virginia Galt, "Forced retirement: a form of age discrimination?"

The Globe and Mail, 22 May 2004.

21 Katherine Harding, "I Loved My Job," *The Globe and Mail*, 16 April 2003.

22 Shawna Richer, "Miner Digs in his Heels over Forced Retirement," *The Globe and Mail*, 16 March 2004.

23 M. Shannon and D. Grierson "Mandatory retirement and older worker employment," *Canadian Journal of Economics* 37:3 (2004), 533.

24 Hélène Mulholland, "Unions Condemn Government Pension Plans in Brighton," *The Guardian*, 13 September 2004.

25 David Neumark, "Age Discrimination in the United States," in Zmira Hornstein, ed., *Outlawing Age Discrimination: Foreign Lessons, UK Choices* (Bristol: The Policy Press, 2001), 62.

26 Till von Wachter, *The End of Mandatory Retirement in the US: Effects on Retirement and Implicit Contracts*, Working Paper No. 49 (University of California, Berkeley: Center for Labor Economics, March 2002), 35.

27 M. Shannon and D. Grierson, *op. cit.*, 533.

28 *The Final Plateau*, 136, 144.

29 Peter Hicks, "The Policy Implications of Aging: A Transformation of National and International Thinking," *Horizons* 6:2 (2003), 6.

30 Schellenberg and Silver, "You can't always get what you want," 2–3.

31 Baum, *The Final Plateau*, 62.

32 Andrea Hamilton, "Senate discusses impact of end of mandatory retirement, 10 years later," *Stanford Report*, 9 April 2003.

33 Ken MacQueen, "The Dawn of a New Old Age."

34 These divisions are highlighted in a recent study of academic staff at the University of Alberta; see Stanley Varnhagen, Myrna Sears, and Guoqiang Zhou, "Report on the Academic Staff Retirement and Pensions Survey," *Academic Technologies for Learning, University of Alberta*, June 2004.

35 See Chapter 10 for an account of the American events.

36 As discussed by John Munro in Chapter 10 and Gillin and Klassen in Chapter 3.

37 David Lepofsky memorably called *McKinney* a "catastrophic downward plunge on the Canadian equality roller coaster ride." David Lepofsky, "The Canadian Judicial Approach to Equality Rights:

Freedom Ride or Roller Coaster?" *Law and Contemporary Problems* 55:1 (1992), 182.

[38] Testimony to the Standing Committee on Administrative Justice Concerning the Ontario Human Rights Code Amendment Act, 1992, Bill 15, December 8, 1992.

[39] See, for example, Ontario Confederation of University Faculty Associations, "Mandatory Retirement Discussion Paper," 3 August 2002. *http://www.ocufa.on.ca/retirement/retire.asp* (Accessed 10 September 2004).

[40] Conversation with Professor Mark Gabbard, President, University of Manitoba Faculty Association, March 2003.

[41] "NDP Won't Change University Rules on Retirement," *Winnipeg Free Press*, 5 December 2002.

[42] York University allows those with low pensions to work past 65 on a reduced load (and salary) basis.

[43] Ontario Confederation of University Faculty Associations, "Mandatory Retirement Discussion Paper."

[44] Orley Ashenfelter and David Card, *Did the Elimination of Mandatory Retirement Affect Faculty Retirement Flows?* National Bureau Economic Research Working Paper No. 8378, July 2001, 959.

[45] This is not surprising, given that the average retirement age in the 1950s for all Canadian workers was near 70.

[46] University of New Brunswick, Faculty of Law: Rand Archive, Newspaper Articles on Ivan C. Rand, *http://www.law.unb.ca/rand/articles.html* (Accessed: 24 February 2004).

[47] *http://www.unb.ca/hr/employees/policies/policy/ general/retire.html* (28 July 2004).

[48] Doreen Duchesne, "More Seniors at Work," 7, 8.

[49] Julie Ann McMullin and Martin Cooke with Rob Downie, "Labour Force Ageing and Skill Shortages in Canada and Ontario," Research Report W/24, Work Network, *Canadian Policy Research Networks*, August 2004, 30.

[50] Ingrid Connidis and Julie McMullin, "Sociological Ambivalence and Family Ties: A Critical Perspective," *Journal of Marriage and the Family*, 64:3 (August 2002), 564.

[51] Friedan, *The Fountain of Age*, New York: Simon and Schuster, 195.

[52] Connidis and McMullin, "Sociological Ambivalence," 564.

[53] Ibid., 565.

[54] Virginia Galt, "Retirees Keen to Return to Work," *The Globe and Mail*, 30 January 2004.

[55] Peter Hicks, "The Policy Implications of Aging," 612.

[56] Cruikshank, *Learning to be Old*, 26.

[57] Pamela Meadows, "Retirement Ages in the UK: A Review of the Literature," *Employment Relations Research Series, no. 18.* (London: UK Department of Trade and Industry, June 2003).

[58] Leanne Elliott, "Revitalizing Universities through Faculty Renewal," Association of Universities and Colleges of Canada, *Research File*, 4:1 (2000), 5.

[59] Ibid.

[60] Marty Thomas and Rosemary A. Venne, "Work and Leisure: A Question of Balance," in David Cheal, ed., *Aging and Demographic Change in Canadian Context* (Toronto: University of Toronto Press, 2003), 213.

[61] For a report illustrating many of these factors, see Duchesne, "More Seniors at Work."

[62] Ibid., 6, 9.

[63] Julie McMullin and Victor Marshall, "Ageism, Age Relations, and Garment Industry Work in Montreal," *The Gerontologist* 41:1 (2001), 112; in Chapter 5, Allan Irving offers three components of ageism, including negative feelings about the old, an ageist ideology, and discriminatory practices.

[64] G. Laws, "Understanding Ageism: Lessons from Feminism and Postmodernism, *The Gerontologist* 35 (1995), 112–18. Quoted from McMullin and Marshall, 113.

[65] Julie Ann McMullin, "Diversity and the State of Sociological Aging Theory," *The Gerontologist* 40:5 (2000), 519, 526. The passages quoted (and re-framed) by McMullin are, respectively, from J.J. Dowd, "The Reification of Age: Age Stratification and the Passing of the Autonomous Subject," *Journal of Aging Studies* 1 (1980), 323 and D. Layder, *Understanding Social Theory* (Thousand Oaks, CA: Sage, 1994), 159.

[66] Martin Lyon Levine, *Age Discrimination and the Mandatory Retirement Controversy* (Baltimore: John Hopkins University Press, 1988), 140.

[67] Peter Laslett, *A Fresh Map of Life: the Emergence of the Third Age* (Cambridge, MA: Harvard University Press, 1991), 152.

[68] This is a paraphrase from the *Dickason* decision; the actual quote is: "[A]ge differs from other grounds of discrimination since everyone of no matter what religion, colour, social origin, nationality or gender becomes older with the passage of time." (1133)

[69] *McKinney v. University of Guelph,* [1990] 3 S.C.R. 229: 312–13.

[70] *Dickason v. Governors of the University of Alberta* (1992), 95 D.L.R. (4th) 439; 2 S.C.R. 1103. *www.lexum.umontreal.ca/csc/scc/en/pub/1992/vol2/html/ 1992scr2_1103.htm*, 1134.

[71] For a detailed analysis of these arguments, see Martin Lyon Levine, *Age Discrimination and the Mandatory Retirement Controversy* (Baltimore: Johns Hopkins University Press, 1988).

[72] McMullin and Marshall, "Ageism," 117, 120, 121.

[73] R. Tamburri, "Rethinking the Rules of Retirement," *University Affairs*, December 2003, 15.

[74] University of British Columbia Senate Ad Hoc Committee on the Academic Implications of Mandatory Retirement at Age 65, 10.

[75] See, for example, a survey conducted by the University of British Columbia, which found that retired faculty are impeded by "lack of office space … or the increasing strains of sharing inadequate space to begin with, among a growing cohort of emeriti." University of British Columbia Senate Ad Hoc Committeee on the Academic Implications of Mandatory Retirement at Age 65, 5.

[76] Canadian Social Sciences and Humanities Federation, "Enrolment and Financial Pressures Will Create New Opportunities for Professors Emeriti," *Perspectives: Special Issue—Congress 2002*, 10.

[77] Fran Abrams and Donald McLeod, "Today, in the Bad Old Days," *The Guardian*, 24 February 2004.

[78] Michael Creal, *Response to President Paul Davenport*, quoted in Canadian Social Sciences and Humanities Federation, "Enrolment and Financial Pressures Will Create New Opportunities for Professors Emeriti," *Perspectives: Special Issue—Congress 2002*, 10.

[79] Y.H. He, A. Colantonio, and A. V. W. Marshall "Later-Life Career Disruption and Self-Related Health: An Analysis of General Social Survey Data," *Canadian Journal on Aging* 22:1 (2003), 55.

[80] David Blane *et al.*, "Inequalities in Quality of Life in Early Old Age," Research Findings: 9 From the Growing Older Programme, *Economic and Social Research Council*, November 2002.

[81] McMullin and Marshall, "Ageism," 121.

[82] Elliott, "Revitalizing Universities," 4, 5.

[83] Douglas Thorpe, "Aging, Language and Culture," in *Aging and Demographic Change in Canadian Context,* ed. David Cheal (Toronto: University of Toronto Press, 2003), 91.

[84] Elliott, "Revitalizing Universities," 5, 7.

[85] Alan Rudrum, personal communication to the author, May 2004. Professor Rudrum is an internationally respected scholar of seventeenth-century English literature.

[86] R.T. Blackburn and J.H. Lawrence, *Faculty at Work: Motivation, Expectation, Satisfaction* (Baltimore: Johns Hopkins University Press, 1995).

[87] Roli Varma, "Permanent Tenure and Academic Freedom in Engineering," *Bulletin of Science, Technology and Society* 2:3 (2001), 197.

[88] Jeff McMaster, "Mandatory retirement," 6.

[89] Tony Atherton, "Academic, journalist still blazing trails at 82: Olive Dickason's ground-breaking life profiled," *The Ottawa Citizen*, 21 May 2003.

[90] *McKinney*, 308.

[91] Ahenfelter and Card, "The Elimination of Mandatory Retirement," 2.

[92] Neumark, "Age Discrimination in the United States," 62. See also von Wachter, "The End of Mandatory Retirement in the US," 36.

[93] Denise K. Magner, "The Imminent Surge in Retirements," *The Chronicle of Higher Education* 46:17 (March 2000), 28.

[94] Jonathan Kesselman, "Mandatory Retirement and Older Workers: Encouraging Longer Working Lives," *C.D. Howe Commentary* No. 200 (June 2004), 12.

[95] Andrea Hamilton, "Senate discusses impact of end of mandatory retirement, 10 years later," 1.

[96] Elliott, "Revitalizing Universities," 3.

[97] Don Savage, Executive Director, Canadian Association of University Teachers, Testimony to the Standing Committee on Administrative

Justice Concerning Ontario Human Rights Code Amendment Act, 1992, Bill 15, 8 December 1992.

[98] Marion Perrin, Executive Director, Ontario Confederation of Faculty Associations, Testimony to the Standing committee on Administrative Justice Concerning Ontario Human Rights Code Amendment Act, 1992, Bill 15, 8 December 1992.

[99] *McKinney*, 284.

[100] Ashenfelter and Card, "The Elimination of Mandatory Retirement," 957-980. Never a friend of either tenure or academic freedom, Boston University attempted in 1999–2000 to parlay loss of mandatory retirement into a weapon to gut tenure. Largely because it would have harmed BU's ability to attract faculty, the effort failed. "It's 10 a.m. Do You Know Where Your Professors Are?" *Chronicle of Higher Education*, 2 February 2001.

[101] Over the past 30 years, the percentage of tenured positions has fallen from 97% to 40% of American faculty. Recently, however, University of Michigan initiated tenure for "faculty specialists" including sessional lecturers and lab technicians. See Ruth Flower, "Tenure Expands," *Academe*, March 2003, *www.aaup.org/publications/Academe/2003/03jf/03jfFTGS.HTM* (19 July 2004).

[102] Ellen Schrecker, "The Incredible Shrinking Faculty: An Interview with Lawrence Poston by Ellen Schrecker,"*Academe,* May–June 2000. *www.aaup.org/publications/Academe/2000/00mj/MJ00Post.htm* (19 July 2004).

[103] Levine, *Age Discrimination,* 100.

[104] A University of British Columbia Senate Report throws a rare light on this turbulent period. "Problems of inadequacies in teaching, research and service, led to resignations, denials of tenure, and terminations. Sometimes the mismatch was so obvious that termination occurred in the middle of candidates' first year." Quoted from Senate Ad Hoc Committee on the Academic Implications of Mandatory Retirement at Age 65, 10.

[105] Kenneth Westhues, *Administrative Mobbing At the University of Toronto-The Trial, Degradation and Dismissal of a Professor During the Presidency of J. Robert S. Prichard* (Lewiston, New York: Edwin Mellen Press, 2004).

[106] Elliott, "Revitalizing Universities," 8.

[107] Ibid., 4.

[108] Caroline Alphonso, "Foreign Hires Sweet Music to Universities," *The Globe and Mail*, 11 May 2004.

[109] Jeffrey Cormier, "Nationalism, Activism, and the Canadian Sociology and Anthropology Community, 1967–1985," *The American Sociologist* 33:1 (Spring 2002), 25.

[110] Ellen Schrecker, "The Incredible Shrinking Faculty."

[111] Betty Freidan, *The Fountain of Age*, 215.

[112] Lorraine T. Dorfman, "Stayers and Leavers: Professors in an Era of No Mandatory Retirement," *Educational Gerontology* 28 (2002), 17.

[113] Samuel Weiss, "Retirement Rules Gone, the Ivory Tower Goes Gray," *The New York Times*, 19 June 2002.

[114] "Center for Retired Professors," *The New York Times*, 13 November 2002.

[115] "Peering Through the Gates of Time," *The New York Times*, 5 March 2002.

[116] "The Light Fantastic," *Business Week*, 2 August 2004.

[117] Tamburri, "Rethinking Mandatory Retirement," 8.

[118] Dr. James Thacker, personal communication, December 2003.

[119] Caroline Alphonso, "Law feeding brain drain," *The Globe and Mail*, 20 February 2004. This article discusses the departure from the University of Toronto by Professor Thomas Pangle, age 60.

[120] Liane Faulder, "The Not-So Golden Handshake: Shades of Grey," *The Edmonton Journal*, 15 February 2004.

[121] Weiss, "Retirement Rules Gone." In Quebec, Concordia University successfully used early retirement packages in the late 1990s to avoid laying off younger faculty (Conversation with Howard Fink, Pension Committee, Concordia University, 31 August 2004.)

[122] Ronald G. Ehrenberg, "Faculty Retirement Policies After the End of Mandatory Retirement," TIAA-CREF Report. Dialogue No. 69 (October 2001); Robert L. Clark, "Keep Senior Faculty or Urge Them to Retire?" *AGB Trusteeship* 11:2 (2003).

[123] Dorfman, "Stayers and Leavers," 16.

[124] Ehrenberg, "Survey of Changes," 6.

[125] Weiss, "Retirement Rules Gone."

[126] Ashenfelter and Card, "Elimination of Mandatory Retirement," 32.

[127] Weiss, "Retirement Rules Gone."

[128] Dorfman, "Stayers and Leavers" 29–30.

[129] "Faculty Grievance: Battle Brewing over Mandatory Retirement at UCC," *Canadian Press*, 19 February 2004.

[130] Stanley Varnhagen, Myrna Sears, and Guoqiang Zhou, "Report on the Academic Staff Retirement and Pensions Survey," 9.

[131] Meyer Brownstone, "Time to Go: Mandatory Retirement, Who Decides?" *University of Toronto Bulletin*, 22 March 2004, 16.

[132] Galt, "Retirees Keen to Return to Work."

[133] Bryon G. Spencer, "Student Enrolment and Faculty Recruitment in Ontario: The Double Cohort, the Baby Boom Echo, and the Aging of University Faculty," *Report to the Ontario Confederation of University Faculty Associations,* October 2001, 38–9.

[134] Ibid., 12.

[135] Howard Fink, "Québec Universities Without Mandatory Retirement: Policies and Experiences," in Peter Russell and Ken Rea, eds., *Redesigning Retirement: Proceedings of a Joint Forum Presented by the University of Toronto Faculty Association and the Retired Academics and Librarians of the University of Toronto*, ed., 5 April 2003, 11–12.

[136] Dorfman, "Stayers and Leavers," 31.

[137] Graham Lowe, quoted in Liane Faulder, "The Not-So-Golden Handshake," *The Edmonton Journal*, 15 February 2004.

[138] Remarks to the Conference and Annual General Meeting of CURAC/ARUCC, Dalhousie University, Halifax, May 2003.

[139] The University of British Columbia Senate Ad Hoc Committee on the Academic Implications of Mandatory Retirement at Age 65, 9–11.

[140] "Renouveler de corps professoral à l'université: des defies importants à mieux cerner," Conseil superieur de l'Éducation, 2002–2003, Québec, December 2003. (Translated by Julie Michaud.)

Chapter 3 The Shifting Judicial Foundation

[1] This line of argument derives from Ian Greene, *The Charter of Rights* (Toronto: James Lorimer & Company Ltd., Publishers 1989).

[2] James Boyd White, *Heracles' Bow: Essays on the Rhetoric and Poetics of the Law* (Madison: The University of Wisconsin Press, 1985), 79. The following line of argument derives from White and from a somewhat different perspective, Paul W. Kahn, *The Cultural Study of Law: Reconstructing Legal Scholarship* (Chicago: The University of Chicago Press, 1999), 36, who asserts, "The rule of law is a social practice: it is a way of being in the world."

[3] *Canadian Human Rights Act* R.S. 1985, c. H-6, Part I.9.(2).

[4] *McKinney v. University of Guelph*, [1990] 3 S.C.R. 229.

[5] White, *Heracles' Bow*, 77, writes: "[T]he most valuable way to talk about the 'meaning' of a text is to ask, among other things, what version of himself or herself it invites its reader to become."

[6] Kahn, *The Cultural Study of Law*, 30, writes: "We must accept the proposition that there is nothing natural about the legal order, that it is a constructed social world that could be constructed differently."

[7] White, *Heracles' Bow*, 95, writes,:"[T]he question, "What does this text mean?"—when properly understood—starts a conversation of a kind we know how to participate in and benefit from. The meaning of a text is not its informational content, nor those statements that any rational person can be forced by logic and fact to make about it, but the cooperative experience it offers its readers."

[8] *Ontario Human Rights Commission v. Borough of Etobicoke,* [1982] 132 D.L.R. (3d) 14.

[9] J. McIntyre, *Etobicoke* at 20.

[10] J. McIntyre, *Etobicoke* at 19–20.

[11] J. McIntyre, *Etobicoke* at 24.

[12] *Re Winnipeg School Division no. 1 and Craton et al.*, [1985] 21 D.L.R. (4th) 1.

[13] Neil Finkelstein and Geoffrey Howard, "Retirement as a Case Study," *The Advocates' Quarterly* 9:2 (1998), 142–59. In part, the optimism about change was fuelled by developments in the United States, where the Age Discrimination in Employment Amendments of 1986 (Public Law 99–592) amended the Age Discrimination in Employment Act of 1967 to remove the existing 70-year upper age limit applicable to employees who are covered under the Act.

[14] M. David Lepofsky, "The Canadian Judicial Approach to Equality Rights: Freedom Ride or Roller Coaster?" *Law and Contemporary Problems* 55:1 (1992), 167–99.

[15] Michael Mandel, *The Charter of Rights and the Legalization of Politics in Canada* (Toronto: Thompson Educational Publishing, 1994). Also see Anne F. Bayefsky and Mary Eberts, "Passage to Retirement: Age Discrimination and the Charter," *Equality Rights and the Canadian Charter of Rights and Freedoms* (Toronto: Carswell, 1985) 231–92.

[16] *McKinney v. University of Guelph*, [1990] 3 S.C.R. 229. *Harrison v. University of British Columbia*, [1990] 77 D.L.R. (4th) 55. *Stoffman v. Vancouver General Hospital*, [1990] 76 D.L.R. (4th) 700. Also decided in 1990, *Douglas/Kwantlen Faculty Association v. Douglas College*, [1990] 77 D.L.R. (4th) 94, was ostensibly a mandatory retirement case, but the legal issues raised were preliminary ones. The arbitrator, in his award, argued that the Charter applied to Douglas College as a Crown agency and that the collective agreement amounted to "law" subject to review under the Charter. These findings formed the basis of the appeal. Neither the arbitrator nor the justices, on appeal, addressed the issues of whether the equality clause was violated or whether the violation was reasonable.

[17] *McKinney v. University of Guelph*, [1987] 46 D.L.R. (4th) 193.

[18] The justices recognized that, based on the Court's decision in *Andrews v. Law Society of British Columbia*, [1989] 1 S.C.R. 143, the restricted definition of age constitutes discrimination under section 15(1) of the Charter.

[19] *McKinney v. University of Guelph*, [1990] 3 S.C.R. 229 at 262.

[20] J. La Forest, *Etobicoke* at 294–5.

[21] J. La Forest, *Etobicoke* at 297.

[22] The following discussion is indebted to Colin G.M. Gibson and Lindsie M. Thomson, "Age Discrimination: An Update," Carman J. Overholt, Course Materials Coordinator, *Employment Law Conference—2003* (Vancouver: The Continuing Legal Education of British Columbia).

[23] J. La Forest, *Etobicoke* at 278.

[24] J. Wilson, *McKinney* at 413 and 415.

[25] Justice L'Heureux-Dubé also noted, at 689: "The adverse effects of mandatory retirement are most painfully felt by the poor."

[26] Justice L'Heureux-Dubé, *McKinney* at 424.

[27] Justice L'Heureux-Dubé, *McKinney* at 430-431.

[28] *Dickason v. University of Alberta,* [1992] 2 S.C.R. 1103.

[29] J. Cory, *Dickason* at 1127.

[30] J. Cory, *Dickason* at 1133-1134.

[31] Shirish P. Chotalia, "The Supreme Court and Mandatory Retirement: Sanctioning the Status Quo," *Constitutional Forum* 4:3 (1993), 67–70.

[32] *R. v. Oakes,* [1986] 1 S.C.R. 103.

[33] The following summaries of *Oakes* and *Irwin Toy* and their implications for the reasoning in *Dickason* are based on Gibson and Thompson, *Age Discrimination.*

[34] *Irwin Toy Ltd. v. Québec (Attorney General),* [1989] 1 S.C.R. 927.

[35] C.J. Dickson, *Irwin Toy* at 993.

[36] J. L'Heureux-Dubé, *Dickason* at 1173.

[37] J. L'Heureux-Dubé, *Dickason* at 1182.

[38] See Lepofsky, "Canadian Judicial Approach to Equality Rights," 167–99.

[39] *Large v. Stratford (City),* [1995] 3 S.C.R. 733.

[40] J. Sopinka, *Large* at para. 22. Note reference citations for Supreme Court judgments change from page numbers to paragraph numbers, following reporting practice on the Supreme Court judgments Web site.

[41] J. Sopinka, *Large* at para. 33.

[42] David Beatty, "The Canadian Conception of Discrimination," *Canadian Labour & Employment Law Journal* 4 (1996), 263–82.

[43] J. Sopinka, *Large* at para. 23.

[44] Beatty, "Canadian Conception of Discrimination," 268–9.

[45] See also M.C. Crane, "Human Rights, *Bone Fide* Occupational Requirements and the Duty to Accommodate: Semantics or Substance?" *Canadian Labour & Employment Law Journal* 4 (1996), 209–32.

[46] *Tétreault-Gadoury v. Canada (Employment and Immigration Commission),* [1991] 2 S.C.R. 22.

[47] J. La Forest, *Tétreault-Gadoury* at 40.

[48] J. La Forest, *Tétreault-Gadoury* at 46.

[49] *British Columbia (Public Service Employee Relations Commission) v. BCGSEU,* [1999] 3 S.C.R. 3.

[50] J. MacLachlin, "Meiorin" at para. 48.

[51] J. MacLachlin, "Meiorin" at para. 50.

[52] See also David Beatty, "Canadian Conception of Discrimination," and Crane, "Human Rights."

[53] J. MacLachlin, "Meiorin" at para. 68.

[54] *Law v. Canada (Minister of Employment and Immigration),* [1999] 1 S.C.R. 4.

[55] J. Iacobucci, *Law* at para. 53, relying on *Rodriguez v. British Columbia (Attorney General),* [1993] 3 S.C.R. 519.

[56] J. Iacobucci, *Law* at para. 73.

[57] J. Iacobucci, *Law* at para. 51.

[58] J. Iacobucci, *Law* at para. 53.

[59] J. La Forest, *Tétreault-Gadoury* at para. 40.

[60] J. Iacobucci, *Law* at para. 80-81.

[61] *GVRD Employees' Union v. GVRD,* [2001] BCCA 435.

[62] J. Prowse, *GVRD* at para. 127.

[63] J. Iacobucci, *Law* at para. 52.

[64] In *Vriend v. Alberta,* 1 S.R.C. [1998] 493 the question before the Supreme Court was whether Alberta's human rights legislation should be interpreted to provide protection against discrimination based on sexual orientation in employment, even though the legislature specifically omitted sexual orientation as a prohibited ground of discrimination. What is of particular importance in this context is that the court reiterated the expansive concept of human rights that is also emerging in other cases, such as *Law* and "Meiorin."

[65] Gavin W. Anderson, "Filling the 'Charter Gap'?": Human Rights Codes in the Private Sector," *Osgoode Hall Law Journal* 33:4 (1995), 749–83.

[66] See Ontario Human Rights Commission, *Time for Action: Advancing Human Rights for Older Ontarians.* June 28, 2001. Toronto: Ontario Human Rights Commission.

[67] Beatty, "Canadian Conception of Discrimination," 281.

[68] Anderson, "Filling the 'Charter Gap'?"; see also Richard Brown, "Human Rights in Employment: of Participation and Compensation," *Canadian Labour & Employment Law Journal* 4 (1996), 283–309.

[69] Crane, "Human Rights," 228–9.

[70] Canadian Human Rights Act Review Panel, *Promoting Equality: A New Vision* (Ottawa: Minister of Justice and the Attorney General of Canada, 2000), 121.

[71] Ibid., 121.

[72] Ontario Human Rights Commission, *Time for Action: Advancing Human Rights for Older Ontarians*, 9. Prior to *Time for Action*, the Ontario Human Rights Commission published a discussion paper, *Discrimination and Age: Human Rights Issues Facing Older Persons in Ontario*, 2000. Subsequent to *Time for Action*, the Commission published its *Policy on Discrimination Against Older Persons Because of Age*, 2002.

[73] Ontario Human Rights Commission, *Time for Action*, 40.

[74] Ian Greene, *The Charter of Rights* (Toronto: James Lorimer & Company Ltd., Publishers, 1989), 228.

[75] Thomas R. Klassen and C.T. Gillin , "The Heavy Hand of the Law: The Canadian Supreme Court and Mandatory Retirement," *Canadian Journal on Aging/La Revue cannadienne du viellissement* 18:2 (1999), 259–76.

[76] See Kirk Makin, "Judicial Activism on Decline," *The Globe and Mail*, (8 January 2005).

[77] See *Law* at paragraph 53.

[78] J. Wilson, dissent in *McKinney* at 391.

[79] Ibid., at 410.

[80] Kahn, *The Cultural Study of Law*, 45, writes, "Law's rule carries forward a past that makes a meaningful claim upon us." Also, see endnote, 4 above.

[81] David Hogg, *Constitutional Law in Canada*, 4th ed. (Toronto: Carswell, 1997), 1284.

[82] Kahn, *The Cultural Study of Law*, 95, writes, "An idea has no force in the world until someone or some community is willing to take a stand with respect to it, i.e., to invest their bodies in the presence of that idea."

Chapter 4 Organized Labour's Predicament

[1] From *The Guinness Dictionary of Quotations for All Occasions*, compiled

by Gareth Sharpe (Middlesex: Guinness Publishing, 1994), 119.

2 Jeremy Seabrook, *A World Growing Old* (Toronto: Between the Lines Press, 2003), 1–6.

3 Some of the data for this paper was obtained from 14 in-depth personal interviews with elected union leaders and staff, human rights commission staff, and observers (academics and labour lawyers) in Toronto, Vancouver, and Ottawa. We are grateful for their assistance and willingness to discuss retirement-related issues.

4 Denis Guest, *The Emergence of Social Security in Canada*, 3rd ed. (Vancouver: University of British Columbia Press, 2003), 75.

5 Ibid., 137–44.

6 Law Commission of Canada, *Law and Relationships between Generations.* (Ottawa: Law Commission of Canada, February 2004), 28–9.

7 Although the Charter of Rights and Freedoms applies only to the actions of government, the various human rights laws that cover all employers must be consistent with the Charter and thus extend its principles to employer-worker relationships.

8 Tom McIntosh, ed., *Federalism, Democracy and Labour Market Policy in Canada* (Montreal: McGill-Queen's University Press, 2000).

9 Neil Bradford, "The Policy Influence of Economic Ideas: Interests, Institutions and Innovation in Canada," *Studies in Political Economy* 59 (Summer 1999), 17–59.

10 See Michael Mandel, *The Charter of Rights and Legalization of Politics in Canada* (Toronto: Thompson, 1994), and Tom McIntosh, *The Post-Charter Labour Regime: The Trilogy and Beyond* (Kingston, Ontario: Industrial Relations Centre, Queen's University, 1989).

11 See Janet L. Hiebert, *Charter Conflicts: What is Parliament's Role?* (Montreal: McGill-Queen's University Press, 2002), and Rainer Knopff and F.L. Morton, *Charter Politics* (Toronto: Nelson Canada, 1992).

12 Knopff and F. L. Morton, *Charter Politics*, 19.

13 Thomas R. Klassen and C.T. Gillin, "The Heavy Hand of the Law: The Canadian Supreme Court and Mandatory Retirement," *Canadian Journal on Aging* 18:2 (1999), 259–76, and C. T. Gillin and Thomas R. Klassen, "Retire Mandatory Retirement," *Policy Options* 21: 6 (2000), 59–62.

14 Lois F. Copperman and Frederick D. Keast, *Adjusting to an Older*

Work Force (New York: Van Nostrand, 1983), 150–1.

15 Statistics Canada Online, Labour Force Survey–3701: Table 282-0051, *www.estat.statcan.ca.*

16 Martin Lyon Levine, *Age Discrimination and the Mandatory Retirement Controversy* (Baltimore: Johns Hopkins University Press, 1988), 65; Melvin K. Bers, *Union Policy and the Older Worker* (Berkeley: Institute of Industrial Relations, University of California, Berkeley, 1957), 83. See also 69–70 for a review of the literature that found a strong opposition to mandatory retirement.

17 Ibid., 83.

18 Morley Gunderson and James Pesando, "The Case for Allowing Mandatory Retirement," *Canadian Public Policy* XIV (1988), 32–9.

19 Buzz Hargrove, "Mandatory Retirement Debate," *The Globe and Mail (online edition)*, 29 December 2003.

20 On this argument, see Neil Guppy, "The Magic of 65: Issue and Evidence in Mandatory Retirement Debate," *Canadian Journal of Aging* 8 (1989), 173–86.

21 Gloria Galloway, "Mandatory retirement may get another look: Many, including Martin, now question if workers should be forced out at 65," *The Globe and Mail*, 26 December 2003, A11.

22 Bob Baldwin [National Director, Social and Economic Policy, Canadian Labour Congress], "Reflections on Mandatory Retirement: Much Ado About Not Very Much," Canadian Labour Congress, *www.clc-ctc.ca* (30 October 2003).

23 Ibid., 2.

24 Ibid., 2.

25 Ibid., 3.

26 Canadian Auto Workers, *Mandatory Retirement Policy Letter #17.* August 13, 1997, *http://www.caw.ca/whoweare/cawpoliciesandstatements/policyletters/ letter_17.asp.*

27 Personal interview, Toronto, 3 March 2004.

28 Pamela Meadows, Retirement Ages in the UK: A Review of the Literature, Employment Relations Research Series No.18 (London: Department of Trade and Industry, 2003).

29 The shift of employers toward defined contribution pensions is central to several recent and current corporate restructuring efforts,

with Air Canada and Stelco being the most high-profile examples.

30 For example, article 39 of the 2001-04 provincial agreement between the Paramedical Professional Bargaining Association and the Health Employers Association of British Columbia reads, "The parties subscribe to the principles of the Human Rights Code of British Columbia."

31 Canadian Association of University Teachers, *CAUT Policy Statement on Retirement,* (Ottawa: November 2002).

32 Canadian Association of University Teachers, *Policy Statement on Retirement and Non-Discrimination on the Basis of Age*, policy #18-1, (Ottawa: April 1994).

33 See Tamra L. Scott, *Human Rights Issues and the Collective Agreement* (Kingston: Queen's University Industrial Relations Centre, 1996) for a summary of some of the major judicial decisions.

34 Personal interview, Toronto, 3 March 2004.

35 Ontario Federation of Labour. *The Right to Retire.* Response by the Ontario Federation of Labour to the Ministry of Labour's Consultation Paper Concerning Mandatory Retirement. (Toronto: September 2004), [italics in original].

36 Linda Briskin and Patricia McDermott, eds., *Women Challenging Unions: Feminism, Democracy and Militancy* (Toronto: University of Toronto Press, 1993); Rosemary Waskett, "The Politics of Difference and Inclusiveness within the Canadian Labour Movement," *Economic and Industrial Democracy* 17 (1996), 587–625.

37 See Martin Lyon Levine, *Age Discrimination and the Mandatory Retirement Controversy,* (Baltimore: Johns Hopkins University Press, 1988), 63–5.

38 Gillian Creese, "Gendering Collective Bargaining: From Men's Rights to Women's Issues," *The Canadian Review of Sociology and Anthropology* 33 (1996), 437–56.

39 Will Herberg, "Bureaucracy and Democracy in Labor Unions," *Antioch Review* 3 (1943), 405–17.

40 Ann Frost, "Union Involvement in Workplace Decision Making: Implications for Union Democracy," *Journal of Labor Research* 21:2 (2000), 266.

41 Robert Michels, *Political Parties: A Sociological Study of the Oligarchical*

Tendencies of Modern Democracy, trans. Eden and Cedar Paul (Glencoe: The Free Press, 1915).

42 Kathryn MacLeod, *The Seniority Principle: Is it Discriminatory?* (Kingston: Queen's University Industrial Relations Centre, 1987).

43 Organization for Economic Co-operation and Development, *Labour Force Statistics 1978–1999* (Paris: OECD, 2000).

44 United States Department of Labor, Labor Force Statistics from the Current Population Survey, News Bulletin: Union Members in 2003. Washington, *stats.bls.gov/ news.release/union2.nr0.htm.*

45 Australian Labour Market Statistics, *Trade Union Membership*. Cat. No. 6105.0. Canberra: Australian Labour Market Statistics, April 2004.

46 New Zealand Council of Trade Unions. *News Release: Union Membership Up—But more Growth Needed*, 19 July 2002, *www.union.org.nz/news.*

47 The unionization rate in Quebec was 37.5 percent in 2003, while for Canada it was 30.5 percent. Statistics Canada, "Fact Sheet on Unionization," *Perspectives on Labour and Income*, 4: 8 (August 2003).

48 From 1997 to 2001, the median age of retirement in Quebec was 59.3, while the median for Canada during this time period was 60.8. Statistics Canada, "Fact Sheet on Retirement," *Perspectives on Labour and Income*, 4:2 (February 2004), 6.

49 In 2003, the *Régie des rentes du Québec* issued a discussion paper on the future of the QPP proposing a direct link between the age at which full benefits are available and years of labour force participation, which would raise the age of eligibility for some, while lowering it for others. For an analysis, see Yvan Guillemette, *Follow Québec's Lead: Removing Disincentives to Work after 60 by Reforming the CPP/QPP*. Commentary #199 (Toronto: C.D. Howe Institute, 2004).

Chapter 5 "Fastened to a Dying Animal"

1 Quoted in Stephen Katz, *Disciplining Old Age: The Formation of Gerontological Knowledge* (Charlottesville: University Press of Virginia, 1996), 90.

2 Ibid., 91.

3 Mary E. Kite and Lisa Smith Wagner, "Attitudes Toward Older Adults," in Todd D. Nelson (ed.), *Ageism* (Cambridge, MA.: The MIT Press, 2002), 131.

[4] William Graebner, *A History of Retirement: The Meaning and Function of an American Institution 1885–1978* (New Haven, Conn.: Yale University Press, 1980).

[5] Chris Phillipson, *Reconstructing Old Age: New Agendas in Social Theory and Practice* (London: Sage, 1998), 35.

[6] Mary E. Kite and Lisa Wagner in, Nelson, *Ageism.*

[7] R.W. Kleemeier, ed., *Ageing and Leisure* (New York: Oxford University Press, 1961), 5.

[8] Mary E. Kite and Lisa Wagner in, Nelson, *Ageism.*

[9] Michel Foucault, *Power/Knowledge: Selected Interviews and Other writings, 1972–77,* edited and translated by Colin Gordon (New York: Pantheon, 1980), 112–13.

[10] Michel Foucault, *The Order of Things: An Archaeology of the Human Sciences* (New York: Vintage, 1973), 158.

[11] Michel Foucault, *Michel Foucault: Beyond Structuralism and Hermeneutics,* eds., Hubert L. Dreyfus and Paul Rabinow (Chicago: University of Chicago Press, 1983), 208.

[12] Katz, *Disciplining Old Age,* 17–18.

[13] Ibid., 19.

[14] Luther H. Martin *et al.*, eds., *Technologies of the Self: A Seminar with Michel Foucault* (Amherst: University of Massachussets Press, 1988), 18. Quoted in Katz, *Disciplining Old Age,* 20.

[15] Scott Bukatman, *Blade Runner* (London: British Film Institute, 1997), 60.

[16] Pamela Lee, *Object To Be Destroyed: The Work of Gordon Matta-Clark* (Cambridge, MA.: The MIT Press, 2001), 184.

[17] Quoted in Katz, 85.

[18] Ibid., 85–86.

[19] Stephen Katz, *Disciplining Old Age,* 94.

[20] Ibid., 95.

[21] Quoted in Ibid., 96.

[22] Raymond Williams, *Modern Tragedy* (Stanford: Stanford University Press, 1966), 100.

[23] Jacquelyn N. Zita, *Body Talk: Philosophical Reflections on Sex and Gender* (New York: Columbia University Press, 1998), 165, 166.

[24] Kelly Oliver, ed., *The Portable Kristeva (updated ed.)* (New York:

Columbia University Press, 2002), 232, 236.

25 I. Young, quoted in Robyn Longhurst, *Bodies: Exploring Fluid Boundaries* (London: Routledge, 2001), 29.

26 Mary Douglas, *Purity and Danger: An Analysis of Concepts of Pollution and Taboo* (London: Routledge, 1966).

27 Longhurst, *Bodies,* 7.

28 Quoted in Longhurst, *Bodies,* 13.

29 Elizabeth Groz, *Sexual Subversions: Three French Feminists* (Sydney: Allen and Unwin, 1989), xiv.

30 Kathy Davis, ed., *Embodied Practices: Feminist Perspectives on the Body* (London: Sage, 1997), 3–4.

31 Ralph Cintron, *Angels' Town: Chero Ways, Gang Life, and Rhetorics of the Everyday* (Boston: Beacon Press, 1997), 212.

32 Ibid., 213.

33 Ibid., 214.

34 Max Horkheimer and Theodor Adorno, *The Dialectic of Enlightenment* (London: New Left Books, 1972, orig. pub. 1947).

35 Charles Altieri, *Postmodernisms Now: Essays on Contemporaneity in the Arts* (University Park, PA.: The Pennsylvania State University Press, 1998), 6.

36 Hal Foster, "Postmodernism in Parallax." *October* 63 (1994), 3.

37 Michel Foucault, *Ethics, Subjectivity and Truth: The Essential Works of Foucault (vol. 1),* ed. Paul Robinow (New York: The New Press, 1997), 261–2.

Chapter 6 Demographic Change

1 Kingsley Davis and P. van den Oever, "Age Relations and Public Policy in Advanced Industrial Societies," *Population and Development Review* 7 (1981), 1–18.

2 David Cheal, ed., *Aging and Demographic Change in Canadian Context* (Toronto: University of Toronto Press, 2003); Peter Hicks, "New Policy Research on Population Aging and Life-Course Flexibility," *Horizons* 6 (2003), 3–6.

3 See Statistics Canada, *Annual Demographic Statistics 2004*. Cat. No. 91-213. (Ottawa: Statistics Canada, 2005).

4 See Statistics Canada, *Profile of the Canadian Population: Where We*

Live. Cat. No. 96F0030. (Ottawa: Statistics Canada, 2002).

5 International Data Base, *United States Bureau of the Census*, International Programs Center. (Washington, DC 2004).

6 Statistics Canada, *Births*. Cat. No. 84F0210XPB. (Ottawa: Statistics Canada, 2003); US Census Bureau, *Global Population Profile 2002*. U.S. Government Printing Office, 2003.

7 Statistics Canada, *Canada's Ethnocultural Portrait: A Changing Mosaic*. Cat. No. 96F0030XIE2001008. (Ottawa: Statistics Canada, 2003d).

8 US Census Bureau, *The U.S. Foreign Born Population: Census 2000*, Current Population Reports, Series P23-206, U.S. Government Printing Office, (Washington, DC).

9 US Census Bureau, *Cumulative Estimates of the Components of Population Change for the United States: April 1, 2000 to July 1, 2003*. Cat. No. NST-EST2003-04. (Washington, DC 2004).

10 "Canada is 30 million, but will it last?" *The Globe and Mail*, 13 May 2002, A1.

11 Population Reference Bureau (2003), *World Population Data Sheet*. Available: *http://www.prb.org* (May 13, 2004).

12 Statistics Canada, *Annual Demographic Statistics 2002*. Cat. No. 91-213. (Ottawa: Statistics Canada, 2003).

13 See the *Review of Demography and its Implications for Economic and Social Policy: Charting Canada's Future*, (Ottawa: Health and Welfare Canada, 1989).

14 See Roderic Beaujot, Suzanne Shiel, and Lorraine Schoel, *Immigration and the Population of Canada: Report Prepared for Immigration Policy Branch, Employment and Immigration* (Ottawa: Employment and Immigration Canada, 1989); Roderic Beaujot and Feng Hou, *Projecting the Visible Minority Population of Canada: the Immigration Component* (Ottawa: Statistics Canada, 1993); Alain Bélanger, *Report on the Demographic Situation in Canada 2000* (Ottawa: Statistics Canada, 2001) Cat. No. 91-209; *Citizenship and Immigration, Statistics 1994* (Ottawa: Minister of Public Works and Government Services Canada) Cat No. MP22-1/1994; *Citizenship and Immigration Statistics 1996* (Ottawa: Citizenship and Immigration Canada); Catalogue no. MP22-1/1997; *Citizenship and Immigration Statistics 1996*. (Ottawa: Minister of Public Works and Government Services Canada). Cat.

No. MP22-1/1996; *Citizenship and Immigration Canada, Statistics 1999* (Ottawa: Minister of Public Works and Government Services Canada). Cat. No. Mp22-1/1999.

[15] See Hervé Le Bras, *The Demographic Impact of Post-War Migration in Selected OECD Countries*. Paper presented at OECD Working Paper on Migration, Paris, 1988.

[16] See Frank T. Denton, Christine H. Feaver, and Byron G. Spencer, *Alternative Pasts, Possible Futures: A "What If" Study of the Effects of Fertility on the Canadian Population and Labour Force*. Paper presented at the meetings of the Federation of Canadian Demographers, Ottawa, December 2001.

[17] Statistics Canada, *Population Projections for Canada and the Provinces and Territories, 1989-2011*. Cat. No. 91-520. (Ottawa: Statistics Canada, 1990).

[18] Jean-Claude Chesnais, *L'inversion de la pyramide des âges en Europe: perspectives et problèmes*. In International Population Conference, 3 (Liège, Belgium: IUSSP, 1989), 53–68.

[19] Statistics Canada, *Population Projections for Canada and the Provinces and Territories, 2000-2026*. Cat. No. 91-520. (Ottawa: Statistics Canada, 2001).

[20] Statistics Canada, "Labour Force Survey, December 2003," *The Daily*, 10 January 2004, 2–5.

[21] See Citizenship and Immigration, *Pursuing Canada's Commitment to Immigration: The Immigration Plan for 2002*. (Ottawa: Citizenship and Immigration Canada, 2001).

[22] Statistics Canada, *The Changing Profile of Canada's Labour Force*. Cat. No. 96F0030XIE2001009. (Ottawa: Statistics Canada, 2003b).

[23] Statistics Canada, "Labour Force Survey, December 2003," *The Daily*, 10 January 2004, 2–5.

[24] Statistics Canada, *Population Projections for Canada and the Provinces and Territories, 2000-2026*. Cat. No. 91-520.(Ottawa: Statistics Canada, 2001).

[25] Montreal Society of Financial Analysts, "Global Ageing and its Impact on the Financial Markets," Presentation by the Office of the Chief Actuary of Canada, 2 April 2003.

[26] See Lars Osberg, "Is It Retirement or Unemployment? Induced

Retirement and Constrained Labour Supply Among Older Workers," *Applied Economics* 25 (1993), 505–19; David Blau, "Labor Force Dynamics of Older Men," *Econometrica* 62 (1994), 117–56; Statistics Canada, *Permanent Layoffs, Quits and Hiring in the Canadian Economy: 1978–1995*. Cat. No. 71-539-XPB (Ottawa: Statistics Canada, 1998); Roman Habtu, "Men 55 and Older: Work or Retire?" *Perspectives on Labour and Income* 15 (2002), 1.

27 Lars Osberg, "Is It Retirement or Unemployment? Constrained Labour Supply and Induced Retirement," *Analytical Studies, Labour Market Activity Survey* (Ottawa: Statistics Canada, 1990).

28 Grant Schellenberg, *Older Workers in the Canadian Labour Market* (Ottawa: The Canadian Seniors Nework, 1994).

29 Geoff Rowe and Huan Nguyen, "Older Workers and the Labour Market," *Perspectives on Labour and Income* 3:12 (2003), 23–6.

30 Economic Council of Canada, *One in Three: Pensions for Canadians to 2030* (Ottawa: Ministry of Supply and Services, 1979); Donald P. Dunlop, *Mandatory Retirement Policy: A Human Rights Dilemma?* (Ottawa: Conference Board of Canada, 1980).

31 Department of Health and Welfare, *Early Retirement: A Preliminary Analysis* (Ottawa: Health and Welfare, 1993).

32 Kelly Cranswick, *Aging and Social Support: General Social Survey*. Cat. No. 89-583-XIE (Ottawa: Statistics Canada, 2003).

33 See Susan Crompton, "One Hundred Years of Health," *Canadian Social Trends* 57 (2000), 2-13; Jiajian Chen and Wayne J. Millar, "Are Recent Cohorts Healthier Than Their Predecessors?" *Health Reports* 11 (2000), 9–23; Sarah Hogan and Jeremy Lise, "Life Expectancy, Health Expectancy and the Life Cycle," *Horizons* 6 (2003), 2.

34 Jean Francois Michaud, M.V. George, and S. Loh, *Projections of Persons with Disabilities*. Cat. No. 91-538. (Ottawa: Statistics Canada, 1996).

35 Statistics Canada, *Life Tables: Canada, Provinces and Territories, 1995-1997*. Cat. No. 84-537. (Ottawa: Statistics Canada, 2002b).

36 Laurent Martel and Alain Bélanger, "An Analysis of the Change in Dependence-Free Life Expectancy in Canada Between 1986 and 1996," *Report on the Demographic Situation in Canada 1998-1999*. Cat. No. 91-209. (Ottawa: Statistics Canada, 1999), 164–86.

37 Frank T. Denton and Byron G. Spencer, "Some Demographic

Consequences of Revising the Definition of "Old Age" to Reflect Future Changes in Life Table Probabilities," *Canadian Journal on Aging* 21 (2002), 3.

[38] Statistics Canada, "Fact Sheet on Retirement," *Perspectives on Labour and Income* 4:9 (2003e), 12–19.

[39] Patrick Kieran, "Early Retirement Trends," *Perspectives on Labour and Income* 2:9 (2001), 5–11.

[40] Susan Crompton and Michael Vickers, "One Hundred Years of Labour Force," *Canadian Social Trends* 57 (2000), 2–13.

[41] Sveinbjorn Blondal and Stefano Scarpetta, *The Retirement Decision in OECD Countries. Economics Department Working Paper Series, no. 202* (Paris: Organization for Economic Cooperation and Development, 1998).

[42] OECD, *Reforms for an Ageing Society* (Paris: Organization for Economic Co-operation and Development, 2000).

[43] David Cheal, ed., *Aging and Demographic Change in Canadian Context* (Toronto: University of Toronto Press, 2003).

[44] Joel Prager, "Aging and Productivity: What Do We Know?" in *Aging and Demographic Change in Canadian Context*, ed. David Cheal (Toronto: University of Toronto Press, 2003), 133–89.

[45] Sarah Hogan and Jeremey Lise, "Life Expectancy, Health Expectancy and the Life Cycle," *Horizons* 6 (2003), 2.

[46] OECD, *Reforms for an Ageing Society.*

Chapter 7 Locating "Mandatory Retirement"

[1] Norene Pupo and Ann Duffy, "Canadian Part-time Work into the Millennium: On the Cusp of Change," *Community, Work and Family* 3:1 (2000), 49–69.

[2] Ann Duffy, Daniel Glenday, and Norene Pupo, "Seniors in the Part-time Labour Force: Issues of Choice and Power," *International Journal of Canadian Studies* 18 (Fall 1998), 67–86.

[3] Throughout this paper, we use the terms "seniors" and "older workers" interchangeably to describe people over the age of 55. Although some analysts restrict the use of senior worker to those age 65 or older, here, references are made to workers aged 55 to 64 as older or senior workers. This is justified on two grounds. First, increasingly in public discourse, age 55 and even 50 are represented

as important markers for the aging worker or consumer. Clearly, the social construction of age is a dynamic aspect of the social order and is currently in upheaval (Irving, Chapter 5). Secondly, this social reorganization of the relationship between age and employment is also reflected in recent research which documents that the average age of retirement for Canadians is *before* age 65. In 2001, only 12 percent of retirees were age 65. See, for example, Statistics Canada, *Labour Force Survey*, Custom Tabulations. (Ottawa: Statistics Canada, 2004a) and Patrick Kieran, "Early retirement trends," *Perspective on Labour and Income* (2001), 7–13.

4 While mandatory retirement is embedded in the law, retirement may be also "constructed" by layoffs, which leave the senior worker with few employment alternatives.

5 For example, the poverty rate for unattached women 65 and older is 45.6 percent, compared to 32.8 percent for comparable males. The National Council of Welfare *Poverty Profile 2001* (Ottawa: Minister of Public Workers and Government Services Canada, 2004), 18. Considerable recent research documents the ongoing economic difficulties of the disabled, visible minorities, and recent immigrants. See, for example, Lesie Harman, "Family Poverty and Economic Struggles,"in N. Mandell and A. Duffy, eds., *Canadian Families: Diversity, Conflict and Change* (Toronto: Thomson Nelson, 2005), 241–75.

6 Jill Quadragno, David MacPherson, Jennifer Reid Keene, and Lori Parham, "Downsizing and the Life-Course Consequences of Job Loss: The Effect of Age and Gender on Employment and Income Security," in Victor W. Marshall, Walter R. Heinz, Helga Kruger, and Anil Verma, eds., *Restructuring Work and the Life Course* (Toronto: University of Toronto Press, 2001).

7 Andrew Jackson, David Robinson, with Bob Baldwin and Cindy Wiggins, *Falling Behind: The State of Working Canada, 2000* (Ottawa: Canadian Centre for Policy Alternatives, 2000), 51, 55.

8 Jon Wells, "Freedom 55, Even 65, Fading Fast for Most Workers." *The Hamilton Spectator*, 10 June 1999, A1–2.

9 Not surprisingly, the possibilities for post-retirement work are related to education. The relationship between post-retirement

employment and education is doubtless complex and may reflect a variety of factors, including relatively low rates of return on hours of work and the physical toll of an unskilled working life. Conversely, "professionals" may be encouraged to continue their employment by the need to pay the high costs of education, a relatively late start in their careers, the ability to pay for assistive services, and the high value on their experience, skills, and presence in the workplace. In short, it appears that highly educated individuals are much better situated to create or maintain post-retirement employment.

10 Doreen Duchesne, "More Seniors at Work," *Perspectives on Labour and Income*. Statistics Canada, Cat. No. 75-001-XIE, 5:2 (February 2004), 3–17, espc. 3.

11 William K. Carroll, *Corporate Power in a Globalizing World* (Toronto: Oxford University Press, 2004), 18.

12 Harry Rudolphs, "Senior drivers just keep on trucking," *Truck News*, 24:4 (April 2004), 1, 18.

13 Not surprisingly, part-time employment is much more likely to be financially feasible in multiple-earner households. In all age groups, women in permanent relationships—married or common law—are more likely than unattached (single) women to work part-time.

14 Statistics Canada, Custom Tabulation 2004.

15 Pupo and Duffy, *op. cit.*, 49–69.

16 Amongst senior women in the public sector, about 74 percent work full-time and 26 percent part-time. This pattern is intensified in the private sector, where one-third of senior women are part-time workers. This prevalence of private part-time employment likely reflects the proliferation of poorly compensated, non-unionized McJobs and the marginalized labour force within this sector.

17 Interestingly, this is a complete reversal of patterns amongst those under 55.

18 Statistics Canada, Custom Tabulation 2004.

19 Snowball sampling refers to a standard sampling procedure employed in qualitative social research. Each research interviewee is asked to suggest the name or names of other individuals who would qualify for the parameters of the research (for example,

retirees who are employed on a part-time basis). In this manner, the size of the sample grows or "snowballs." Although the resultant sample is not representative and does not lend itself to broad generalization, it is useful in identifying key themes and issues as well as giving voice to social experiences. Any identifying comments have been deleted or altered.

20 Gangaram Singh and Anil Verma, "Is There Life after Career Employment? Labour-Market Experience of Early Retirees," in Victor W. Marshall, Walter R. Heinz, Helga Krueger, and Anil Verma, eds., *Restructuring Work and the Life Course* (Toronto: University of Toronto Press, 2001).

21 Daniel C. Feldman and Kim Seongsu, "Bridge Employment During Retirement: A Field Study of Individual and Organizational Experiences with Post-Retirement Employment," *Human Resource Planning* 23:1 (2000), 14–26.

22 As documented extensively, women do tend to be under-represented in general in the post-retirement workforce.

23 Wendy Pyper and Philip Giles, "Approaching Retirement," *Perspectives on Labour and Income* (Winter 2002), 9–16.

24 Dave Gower, "Income Transition Upon Retirement," *Perspectives on Labour and Income*. Statistics Canada, Cat. No. 75-001-XPE, 10:4 (Winter 1998), 18–23.

25 This finding is, of course, in keeping with the analysts who argue that for many workers work has become home. See Arlie Russell Hochschild, *The Time Bind: When Work Becomes Home and Home Becomes Work* (New York: Henry Holt and Company, 1997).

26 See, for example, Anne Wilson Schaef and Diane Fassel, *The Addictive Organization* (San Francisco: Harper and Row Publishers, 1988); and Juliet B. Schor, *The Overworked American: The Unexpected Decline of Leisure* (New York: Basic Books, 1991).

27 Ann Robertson, "'I Saw The Handwriting on the Wall': Shades of Meaning in Reasons for Early Retirement," *Journal of Aging Studies* 14:1 (2000), 63–79.

28 An important, gendered, exception to this pattern is the provision of child care by grandparents, notably grandmothers. Engaging in some form of child care may mean a deferral of "retirement" sta-

tus. Women are less likely to consider themselves retired if they have responsibilities for child care—see Namkee G. Choi, "Self-Defined Retirement Status and Engagement in Paid Work Among Older Working-Age Women: Comparison between Childless Women and Mothers," *Sociological Inquiry*, 72:1 (Winter 2002), 43–71. Of course, for some women, the responsibility of caring for grandchildren may actually necessitate retirement from paid employment—see Emma Dentinger and Marin Clarkberg, "Informal Caregiving and Retirement Timing Among Men and Women," *Journal of Family Issues* 23:7 (October 2002), 857–79.

[29] Neena Chowdhury, "Self-Esteem Drops as People Age, Study Finds," *Toronto Star*, 22 March 2004, E7.

[30] Chowdhury, *op. cit.*; Janet Fast and Judith Frederick, "The Time of Our Lives: Juggling Work and Leisure Over the Life Cycle," Statistics Canada, Cat. No. 89-584-MIE-Number 4 (1998); William Graebner, "Retirement and the Origins of Age Discrimination," in Meredith Minkler and Carroll L. Estes, eds., *Readings in the Political Economy of Aging* (Farmingdale, New York: Baywood Publishing Company, 1984), 177–96; Laurie Russell Hatch, *Beyond Gender Differences: Adaptation to Aging in Life Course Perspective* (Amityville, New York: Baywood Publishing Company, 2000); Phyllis Moen, "A Life-Course Approach to Retirement and Social Integration," in Karl Pillemer *et al.*, eds., *Social Integration in the Second Half of Life* (Baltimore: The Johns Hopkins University Press, 2000), 75–107.

[31] Robert A. Stebbins, "The Extraprofessional Life: Leisure, Retirement and Unemployment," *Current Sociology* 48:1 (January 2000), 1–18.

[32] *Toronto Star*, 2 March 2004, D5.

[33] This flexible approach must, of course, weigh any competing interests between employers, employees, and union representatives. In the late 1990s, employers and unions at Air Canada, Canadian Airlines, and the Canadian federal government introduced a phased-in retirement program that allowed workers to retire at age 55, begin to collect pensions and to continue to work on a part-time basis. Ideally, benefits would accrue to all participants. "Retirees" would retain a portion of their income, would continue

with their work-related social networks and would retain seniority scheduling advantages. At the same time, they would have more free time to experiment with new life directions. New workers would benefit as more full- and part-time jobs were created to partially replace the retirees. Employers would benefit by retaining the experience and knowledge of senior workers but at reduced costs. In this particular instance, conflict emerged between the senior part-time retirees and the regular part-timers, who found that they were being bumped by the seniors in terms of scheduling and vacations. See Vanessa Lu, "Retirement plan aids workers of all ages," *Toronto Star*, 7 December 1998, B1, B4. This example speaks to the complexity of issues at hand.

34 Sara E. Rix, "Restructuring Work in an Aging America: What Role for Public Policy?" in *Restructuring Work and the Life Course.*

35 Anders Hayden, *Sharing the Work, Sparing the Planet: Work Time, Consumption and Ecology* (Toronto: Between the Lines Press, 1999).

Chapter 8 Not as Simple as it Seems

1 Hon. James K. Bartleman, "Speech from the Throne," April 30, 2003 *hansardindex.ontla.on.ca/hansardeissue/37-4/l001.htm* (October 15, 2004).

2 Heather Scoffield, "Rethink CPP's Age-65 Rule," *The Globe and Mail*, 21 April 2004, A1.

3 Morley Gunderson, *Banning Mandatory Retirement: Throwing the Baby Out with the Bathwater* (Toronto: C.D. Howe Institute, 2004), 1.

4 See Morley Gunderson and James Pesando, "The Case for Allowing Mandatory Retirement," *Canadian Public Policy* 14:1 (1988), 33; see also Rafael Gomez, Morley Gunderson, and Andrew Luchak, "Mandatory Retirement: A Constraint in Transitions to Retirement?" *Employee Relations* 24:4 (2002), 403–22, as well as references cited therein. More recent evidence from a Hewitt Associates survey [Hewitt Associates, *Mandatory Retirement: Current Practice and Future Directions Survey Report* (Toronto: Hewitt Associates, 2003)], indicates that 52 percent of Canadian organizations of 100 and more employees had a company-wide mandatory retirement policy, with some others having such a policy for part of

their workforce. As Kesselman points out, however, the restriction to larger firms likely overstates the extent of mandatory retirement for the whole workforce; see Jonathan Kesselman, *Time to Retire Mandatory Retirement.* Paper prepared for C.D. Howe Institute. (Toronto: C.D. Howe Institute, 2004), 9.

5 The numbers from which these calculations are based are from Gunderson, *Banning Mandatory Retirement*, 2.

6 Empirical evidence in Frank Reid, "Economic Aspects of Mandatory Retirement: The Canadian Experience," *Relations industrielles/industrial relations* 43:1 (1988), 101-113 and Michael Shannon and D. Grierson, "Mandatory Retirement Bans and Older Worker Employment," *Canadian Journal of Economics* 36:3 (2004), 528–51, suggests that the bans on mandatory retirement in Manitoba and Quebec did not substantially increase the labour force participation of older workers. However, the very year that mandatory retirement was banned in Quebec, that province also introduced other changes in the pension system that would discourage labour force participation, thereby potentially offsetting any effect of banning mandatory retirement. As stated by Francois Vaillancourt, "The Quebec Pension Plan," in Paul Boothe, ed., *A Separate Pension Plan for Alberta* (Edmonton: University of Alberta Press, 2000), 32, in discussing the Quebec ban on mandatory retirement: "this incentive to work longer was counteracted by the introduction in 1984 of the right to receive an actuarially reduced (0.5 percent per month of anticipation) [Régime des Rents du Quebec] between the ages of 60 and 64 prior to the standard age of retirement, 65. Other provinces also chose to abolish the compulsory age of retirement but did not have the flexibility to encourage early retirement since they were members of the CPP [Canada Pension Plan]." Furthermore, aggregate data indicates that while the labour force participation rate of older workers was declining steadily in both Canada and the United States from 1976 to 1986, after 1986 (the year mandatory retirement was banned in the United States), the labour force participation rate increased sharply in the United States but continued to decline in Canada; see Morley Gunderson, Douglas Hyatt, and James Pesando, "Public Pension Plans in the United States and

Canada," in William Alpert and Stephen Woodbury, eds., *Employee Benefits and Labor Markets in Canada and the United States* (Kalamazoo, MI: Upjohn Institute for Employment Research, 2000), 404; Rafael Gomez, Morley Gunderson, and Andrew Luchak, "Mandatory Retirement: Lessons from Canada," University of Toronto Working Paper, 2003. As well, recent evidence from the United States suggests that the ban on mandatory retirement did have a substantial effect on postponing the retirement of university professors; see Orley Ashenfelter and David Card, "Did the Elimination of Mandatory Retirement Affect Faculty Retirement?" *American Economic Review* 92: 4 (2003), 957–80 and Robert Clark, Linda Ghent, and Juanita Kreps, "Faculty Retirement in Three North Carolina Universities," in Robert Clark and P. Brett Hammond, eds., *To Retire or Not? Retirement Policy and Practice in Higher Education,* (Philadelphia: University of Philadelphia Press, 2001), 21–38.

7 This unfortunate acceptance is discussed in Ontario Human Rights Commission, *Discrimination and Age: Human Rights Issues Facing Older Persons in Ontario* (Toronto: Ontario Human Rights Commission, 2000) and Ontario Human Rights Commission, *Time for Action: Advancing Human Rights for Older Ontarians* (Toronto: Ontario Human Rights Commission, 2001).

8 David MacGregor, Chapter 2 in this volume.

9 This argument is articulated in Kesselman in Chapter 9 and Michael Krashinsky, "The Case for Eliminating Mandatory Retirement: Why Economics and Human Rights Need Not Conflict," *Canadian Public Policy* 14:1 (1988), 40–51.

10 This is one of the reasons that comparisons between Canada and the US are not appropriate. While mandatory retirement in the US is banned, retirement is discouraged by the clawbacks in the Social Security system (the equivalent of our Canada Pension Plan), as well as the higher age of eligibility for Social Security.

11 Heather Scoffield, "Rethink CPP's Age-65 Rule," *The Globe and Mail,* 21 April 2004, A1.

12 Pesando and Gunderson, "Allowing Mandatory Retirement," 1988, 259.

[13] An exception is Jonathan Kesselman, *Mandatory Retirement of Older Workers: Encouraging Longer Working Lives* (Toronto: C.D. Howe Institute, 2004), and also Chapter 9, where he explicitly links a ban on mandatory retirement to possible increases in the age of entitlement for tax-funded programs for older persons. Also cited in Ken MacQueen, "Dawn of a New Age," *Macleans*, 24 May 2004, 70.

[14] David Croll, *Retirement Without Tears: Report of the Special Senate Committee on Retirement Age Policies* (Ottawa: Supply and Services, 1979), 26.

[15] Lynn McDonald, "Retirement for the Rich and Retirement for the Poor: From Social Security to Social Welfare," *Canadian Journal on Aging* 14:3 (1995), 447.

[16] C.T. Gillin and Thomas Klassen, "Retire Mandatory Retirement," *Policy Options* 21:6 (2000), 59–62; Klassen and Gillin, "The Heavy Hand of the Law: The Canadian Supreme Court and Mandatory Retirement," *Canadian Journal on Aging/La Revue cannadienne du viellissement* 18:2 (1999), 261, 268.

[17] James Pesando, Morley Gunderson, and James McLaren, "Pension Benefits and Male-Female Wage Differentials," *Canadian Journal of Economics* 24:3 (1991), 536–50.

[18] These rationales for mandatory retirement are outlined, for example, in Morley Gunderson and James Pesando, "Eliminating Mandatory Retirement: Economics and Human Rights," *Canadian Public Policy* 6:2 (1980), 352–60; Gunderson and Pesando, "Allowing Mandatory Retirement," 1988; and Gomez, Gunderson and Luchak, "Mandatory Retirement," 2003.

[19] As stated by MacGregor in Chapter 2 in the context of universities, "Faculty renewal may be the consummate ageist term, suggesting that older faculty resemble a condemned neighbourhood in a rotted city core."

[20] The lump-of-labour fallacy at the aggregate level is acknowledged by economists who are in favour of allowing mandatory retirement. See Gunderson and Pesando, "Eliminating Mandatory Retirement," 1980, 358 and Pesando, *The Elimination of Mandatory Retirement,* 1979, 21, as well as those opposed to allowing it. See Kesselman, 2004 and Chapter 9, and Krashinsky, "The Case for Eliminating Mandatory Retirement," 1988, 47.

21 Sandra E. Martin, "Mandatory Retirement Far From Decided," *National Post*, 31 March 2004, FP12.

22 Implications of banning mandatory retirement for a range of personnel functions are discussed in Morley Gunderson, "Mandatory Retirement and Personnel Policies," *Columbia Journal of World Business* 28:2 (1983), 8–15.

23 As indicated by Wayne Samuelson, President of the Ontario Federation of Labour, "If age isn't the criterion for leaving the workforce, performance will be. Older workers with higher insurance and disability costs will be fired at the first dip in productivity—an uglier end to one's working life." Wayne Samuelson, quoted in Patricia Treble, "Over the Hill? What Hill?" *Macleans,* 24 May 2004, 70.

24 It is important to emphasize that neither this rationale for mandatory retirement nor any of the rationales discussed so far are premised on any notion that the productivity of older workers declines with age. The general literature suggests that there is no clear relationship between productivity and age. Reviews of that evidence are given in Judith Hellerstein, David Neumark, and Kenneth Troske, "Wages, Productivity and Worker Characteristics: Evidence from Plant-Level Production Functions and Wage Equations," *Journal of Labor Economics* 17:3 (1999), 409–46; Mary Jablonski, Kent Kunze, and Larry Rosenblum, "Productivity, Age and Labour Composition Changes in the US Work Force," in Irving Bluestone, Rhonda Montgomery and John Owen, eds., *The Aging of the American Workforce* (Detroit: Wayne State University Press, 1990), 304–38; Richard Posner, *Aging and Old Age* (Chicago: University of Chicago Press, 1995), 66–98, 156–201; Josef Richter, "Economic Aspects of Aging: Review of the Literature," in George Stolnitz , ed., *Demographic Causes and Economic Consequences of Population Aging* (New York: UN Economic Commission for Europe, 1992), 171–86; Anthony Sterns, Harvey Sterns, and Lisa Hollis, "The Productivity and Functional Limitations of Older Adult Workers," in William Crown, ed., *Handbook on Employment and the Elderly* (Westport, Conn.: Greenwood Press, 1996), 276–303.

25 Edward Lazear, "Why Is There Mandatory Retirement?" *Journal of Political Economy* 87:6 (1979), 1261–84.

[26] Empirical evidence on the prevalence of deferred wages is discussed, for example, in Laurence Kotlikoff and Jagadeesh Gokhale, "Estimating a Firm's Age-Productivity Profile Using the Present Value of Earnings," *Quarterly Journal of Economics* 107:4 (1992), 1215–42; Lazear, 1979, and Canice Prendergast, "The Provision of Incentives in Firms," *Journal of Economic Literature* 37 (1999), 7–63.

[27] James Pesando and Morley Gunderson, "Retirement Incentives Contained in Occupational Pension Plans and their Implications for the Mandatory Retirement Debate," Canadian Journal of Economics 88 (1988), cite US evidence on pension benefit accruals and provide evidence for representative pension plans in Canada that such pension benefit accruals can easily amount to 20 percent of the compensation of workers between the ages of 45 and 65. As such, even if deferred *wages* do not exist—and our interpretation is that they do—deferred total *compensation* in such situations with mandatory retirement invariably exists in the form of age-related benefits and especially pension benefit accruals.

[28] This point is emphasized in Lorne Carmichael, "Firm-Specific Human Capital and Promotion Ladders," *Bell Journal of Economics* 14 (1983), 251–61.

[29] Survey evidence indicating that *employees prefer* deferred compensation is given in Robert Frank and Robert Hutchens, "Wages, Seniority and the Demand for Rising Consumption Profiles," *Journal of Economic Behavior and Organization* 21 (1993), 251-276 and in George Loewenstein and Nachum Sicherman, "Do Workers Prefer Increasing Wage Profiles?" *Journal of Labor Economics* 9 (1991), 67-84.

[30] The legislative background for mandatory retirement is outlined in Chapter 3 by Gillin and Klassen; Morley Gunderson, "Age Discrimination in Employment in Canada," *Contemporary Economic Policy* 21:3 (2003), 318–28; Klassen and Gillin, "The Heavy Hand of the Law," *op. cit.*; Ontario Human Rights Commission, 2000; and R. Zinn and P. Brethour, *Law of Human Rights in Canada: Practices and Procedures* (Aurora: Canada Law Book, 1999), s.3, 1–4.

[31] *Parent v. The Gazette,* 81 D.L.R. (4th) 689 (A.Q.) (1991).

[32] In the early 1980s, Alberta eliminated mandatory retirement for its

civil servants, but allowed it for other areas like higher education, as quoted from MacGregor, Chapter 2.

33 *McLaren v. Pacific Coast Savings Credit Union,* British Columbia Court of Appeals, 186 (2000).

34 *Greater Vancouver Regional District Employees' Union v. Greater Vancouver Regional District.* British Columbia Court of Appeals, 435 (2001).

35 Case law on BFORs is discussed by Gillin and Klassen in Chapter 3; Klassen and Gillin, "The Heavy Hand of the Law," 1999; and the Ontario Human Rights Commission, *Discrimination and Age,* 2000.

36 Re-evaluating such barriers, for example, is recommended in both Morley Gunderson, *Flexible Retirement as an Alternative to 65 and Out* (Toronto: C.D. Howe Institute, 1998), who favours allowing mandatory retirement, and Kesselman in Chapter 9, who favours banning mandatory retirement.

37 US evidence in David Neumark and Wendy Stock, "Age Discrimination Laws and Labor Market Efficiency," *Journal of Political Economy* 107:5 (1999), 1081–125, for example, indicates that such protection against age discrimination can be important in facilitating deferred compensation, since employees have more security in receiving their deferred compensation.

38 If the age cap is removed, it is important that renegotiations of contracts under different terms and conditions of employment be allowed. This could include lower pay, for example, to offset higher age-related benefit costs.

Chapter 9 Challenging the Economic Assumptions

1 Wassily Leontief, "Theoretical Assumptions and Nonobserved Facts," *American Economic Review* 61:1 (March 1971), 1–7, at 2.

2 Manitoba and Quebec banned mandatory retirement in 1982 and 1983, respectively. At the time of writing, British Columbia, Newfoundland, Ontario, and Saskatchewan did not protect against age discrimination in employment for persons aged 65-plus, thus allowing employers to set retirement ages. The other provinces protect older workers against age discrimination with the exception of mandatory retirement. See Michel Fourzly and Marc Gervais, *Collective Agreements and Older Workers in Canada* (Ottawa: Labour

Program, Human Resources Development Canada, 2002), 156; Morley Gunderson, "Age Discrimination in Employment in Canada," *Contemporary Economic Policy* 21:3 (2003), 318–28; and Gillin and Klassen (Chapter 3).

3 Some analysts have argued that contractual mandatory retirement should continue to be permitted but that there should be no age limit on the prohibition of *other* forms of employment discrimination. See Gunderson and Hyatt in Chapter 8; Gunderson, "Age Discrimination"; and Morley Gunderson, "Banning Mandatory Retirement: Throwing Out the Baby with the Bathwater," *C.D. Howe Institute Backgrounder* 79 (Toronto: C.D. Howe Institute, 2004).

4 See Edward P. Lazear, "Why Is There Mandatory Retirement?" *Journal of Political Economy* 87:6 (1979), 1261–84, and James E. Pesando, *The Elimination of Mandatory Retirement: An Economic Perspective* (Toronto: Ontario Economic Council, 1979).

5 Note that wage rates below productivity levels will also be accepted by workers in their earlier years if they are gaining general human capital skills readily transferable to employment elsewhere.

6 Pesando, *The Elimination of Mandatory Retirement*, 4.

7 Morley Gunderson and James Pesando, "The Case for Allowing Mandatory Retirement," *Canadian Public Policy/Analyse de politiques* 14:1 (1988), 32–9.

8 *McKinney v. University of Guelph*, [1990] 3 S.C.R. 229, at 294–295, 236; henceforth cited as *McKinney*.

9 "Draft Policy #52: Retirement and Post-Retirement," advanced by the UBC administration for discussion, *Faculty Focus: Newsletter of the Faculty Association of the University of British Columbia* 36:7 (2003), 10. Note that UBC was a party to one of the cases considered alongside *McKinney*.

10 In this section I question the implicit assumption that all workers have the same preferences about when to retire, so that a collective decision will satisfy all their needs. In a later section, I introduce heterogeneity in the productivity levels of older workers as well as heterogeneity in their desired retirement age. Early Canadian advocates of mandatory retirement cited possible differences of preferred retirement ages between younger and older members of

a union but asserted that the union would properly mediate. See Gunderson and Pesando.

11 Gunderson, "Banning Mandatory Retirement," 2–3.

12 Lazear, "Why Is There Mandatory Retirement?" 1281. Note that these figures relate to a period prior to the abolition of mandatory retirement in the United States.

13 Gunderson and Pesando, "Allowing Mandatory Retirement," 33.

14 This point was acknowledged in the Supreme Court's majority judgment by reference to situations "... where a collective agreement may not really find favour with individual employees subject to discrimination." *McKinney* at 233.

15 Lazear, "Why Is There Mandatory Retirement?" 1282.

16 A classic example of the female worker entering a job late in life was that of Olive Dickason, who completed her doctorate in 1977 at age 57, worked as a professor at the University of Alberta to age 65 in 1985 and continued working while her legal challenge of mandatory retirement wended its way through the courts, ending in 1992 with an adverse ruling by the Supreme Court of Canada in the case *Dickason v. University of Alberta,* [1992] 2 S.C.R. 1103. Dickason, who was part-Métis, pioneered the field of Canadian Aboriginal history.

17 See Michael Krashinsky, "The Case for Eliminating Mandatory Retirement: Why Economics and Human Rights Need Not Conflict," *Canadian Public Policy/Analyse de politiques* 14:1 (1988), 40–51.

18 See P.A. Diamond, "A Framework for Social Security Analysis," *Journal of Public Economics* 8:3 (1977), 275–98 and Peter Diamond and Botond Köszegi, "Quasi-Hyperbolic Discounting and Retirement," *Journal of Public Economics* 87: 9-10 (2003), 1839–72.

19 Statistics Canada. *Labour Force Update: Older Workers,* Cat. No. 71-005-XPB. Ottawa: Statistics Canada, Spring 1998, 22.

20 Gunderson, "Banning Mandatory Retirement," 3.

21 Lazear, "Why Is There Mandatory Retirement?" 1263. Gunderson and Hyatt in Chapter 8 similarly reject this common argument.

22 Quoted in Krashinsky, "The Case for Eliminating Mandatory Retirement," 50. See also the earlier quotation from the University of British Columbia about "faculty renewal," and Chapter 4 by Klassen with Forgione.

[23] Pesando, *The Elimination of Mandatory Retirement,* 23.

[24] Ontario Human Rights Commission, *Time for Action: Advancing Human Rights for Older Ontarians* (Toronto: 2001), 37. Also see the examples cited by MacGregor in Chapter 2.

[25] These statistics are taken from Human Resources Development Canada, *Challenges of an Aging Workforce: An Overview of the Issue* (Ottawa, 2002).

[26] These figures were provided in Gunderson, "Banning Mandatory Retirement," 2. The half coverage by mandatory retirement was confirmed in a recent survey by Hewitt Associates, *Mandatory Retirement: Current Practice and Future Directions Survey Report* (Toronto: 2003), 5, which found that 52 percent of Canadian companies had an organization-wide mandatory retirement policy (and a few more percent with such policies on an incomplete basis), though the survey's exclusion of firms with fewer than 100 employees likely overstates the figure for the overall labour force.

[27] M. Shannon and D. Grierson, "Mandatory Retirement and Older Worker Employment," *Canadian Journal of Economics* 37:3 (2004), 528–51; Frank Reid, "Economic Aspects of Mandatory Retirement: The Canadian Experience," *Relations Industrielles* 43:1 (1988), 101–13.

[28] Orley Ashenfelter and David Card, "Did the Elimination of Mandatory Retirement Affect Faculty Retirement?," *American Economic Review* 92:4 (2002), 957–80.

[29] For discussion of these issues, see Morley Gunderson, "Flexible Retirement as an Alternative to 65 and Out," *C.D. Howe Institute Commentary 106* (Toronto: C.D. Howe Institute, May 1998) and Jonathan R. Kesselman, "Mandatory Retirement and Older Workers: Encouraging Longer Working Lives," *C.D. Howe Institute Commentary 200* (Toronto: C.D. Howe Institute, June 2004).

[30] However, in the formal economic theory of contractual mandatory retirement, it is not necessary that the productivity of older workers decline, but simply that their compensation exceeds their productivity.

[31] For various statements of this view, see Lazear, "Why Is There Mandatory Retirement?", Gunderson and Pesando, "Allowing Mandatory Retirement," and Gunderson, "Age Discrimination,"

and "Banning Mandatory Retirement." Here, I consider the productivity of older workers as an exogenous matter of health status, while in the next section, I consider the worker incentive aspects of productivity; both are subsumed in the economic model.

32 For a review of studies, see Anthony A. Sterns, Harvey L. Sterns, and Lisa A. Hollis, "The Productivity and Functional Limitations of Older Adult Workers," in William H. Crown, ed., *Handbook on Employment and the Elderly* (Westport, CT: Greenwood Press, 1996), 276–303 and Judith K. Hellerstein, David Neumark, and Kenneth R. Troske, "Wages, Productivity, and Worker Characteristics: Evidence from Plant-Level Functions and Wage Equations," *Journal of Labor Economics*, 17:3 (1999), 409–46. Hellerstein *et al.* find that the productivity of workers continues to rise with age, even for the age group 55 and over, though not with statistical significance.

33 Josef Richter, "Economic Aspects of Aging: Review of the Literature," in George J. Stolnitz, ed., *Demographic Causes and Economic Consequences of Population Aging: Europe and North America* (New York: United Nations Economic Commission for Europe, 1992), 171–86, at 181.

34 For a textbook presentation of the formal economic model, see Dwayne Benjamin, Morley Gunderson, and W. Craig Riddell, *Labour Market Economics: Theory, Evidence, and Policy in Canada*, 5th ed. (Toronto: McGraw-Hill Ryerson, 2002), 47–9.

35 Labour Canada, *An Industrial Relations Perspective on Mandatory Retirement* (Ottawa: Labour Canada, 1985), Annex, 5.

36 For supportive findings, see Laurence J. Kotlikoff and Jagadeesh Gokhale, "Estimating a Firm's Age-Productivity Profile Using the Present Value of Workers' Earnings," *Quarterly Journal of Economics* 107:4 (1992), 1215–42 and Canice Prendergast, "The Provision of Incentives in Firms," *Journal of Economic Literature,* 37:1(1999), 7–63, at 47–49. For opposing findings, see Hellerstein, Neumark, and Troske, "Wages, Productivity, and Worker Characteristics."

37 Gunderson and Pesando, "Allowing Mandatory Retirement," 36 (emphasis added). This assertion continues to be made, as in a recent labour economics textbook: "[S]uch a contractual arrangement requires a termination date for it to exist. Otherwise,

employers run the risk of paying wages in excess of productivity for an indefinite period.... In such circumstances a contractual arrangement involving deferred wages could not persist." See Benjamin, Gunderson, and Riddell, *Labour Market Economics,* 407.

38 A survey by the Conference Board of Canada found that "those few employees presently aged 65 who elect to continue working will, in general, leave their employment prior to age 70 and will, on average, probably work for a period of three years beyond age 65." Donald P. Dunlop, *Mandatory Retirement: A Human Rights Dilemma?* (Ottawa: The Conference Board of Canada, 1980), 40.

39 David Neumark and Wendy A. Stock, "Age Discrimination Laws and Labor Market Efficiency," *Journal of Political Economy* 107:5 (1999), 1081–1125.

40 The number of Old Age Security Pension (OAS) beneficiaries is projected to rise from 3.8 million in 2001 to 8.4 million in 2030 and the number of GIS beneficiaries from 1.4 million to 2.2 million. Annual OAS expenditures are projected to rise from $19.5 billion in 2001 to $89.5 billion in 2030 and annual GIS expenditures from $5.3 billion to $18.0 billion. Office of the Chief Actuary, Office of the Superintendent of Financial Institutions, *Actuarial Report* (5th) *on the Old Age Security Program as at 31 December 2000*, tabled before Parliament on 19 June 2002, 25, 29.

41 Several provinces pay supplementary cash benefits to lower-income GIS recipients, and their budgets are also impacted by workers who are forced to retire.

42 Stephen E. Snyder and William N. Evans, "The Impact of Income on Mortality: Evidence from the Social Security Notch," Working Paper 9197 (Cambridge, MA: National Bureau of Economic Research, 2002) and David Blane *et al.*, "Inequalities in Quality of Life in Early Old Age," *Research Findings: 9*, ESRC Growing Older Programme (Department of Sociological Studies, University of Sheffield, UK, 2002).

43 These figures are based on projections that assume the continuation of existing trends in immigration, birth, death, and participation rates. See David Baxter and Andrew Ramlo, "Changing People, Changing Participation: Demographic and

Behavioral Trends as a Context for the Future of the Canada Pension Plan, 2001 to 2051," paper presented at the Third Inter-Disciplinary Seminar of the Office of the Chief Actuary (Ottawa: Office of the Superintendent of Financial Institutions, 2003), 21.

44 William B.P. Robson, *Aging Populations and the Workforce: Challenges for Employers*, British-North American Committee, available at *www.cdhowe.org/pdf/ BNAC_Aging_Populations.pdf*, 7. See also the discussion of population aging by Kerr and Beaujot in Chapter 6.

45 For evidence on and analysis of this issue in a cross-country context, see chapter 5 of Organisation for Economic Co-operation and Development, *Ageing and Income: Financial Resources and Retirement in 9 OECD Countries* (Paris: OECD, Social Issues, 2001).

46 These figures were compiled from the Labour Force Survey by Régie des rentes du Quebec, *Adapting the Pension Plan to Québec's New Realities*, Working Paper (Quebec City: Ministry of Employment, Social Solidarity and Family Welfare, October 2003), 23. To the extent that longevity is correlated with educational level, the figures for persons aged 65–74 in 2020 are likely to be even more favourable than for those aged 55–64 in 2010. For Census figures on educational attainment by seniors in 2001, see Doreen Duchesne, "More Seniors at Work," *Perspectives on Labour and Income* 5:2 (2004), 5–17, at 6–8.

47 Also see the deeper examination of issues of mandatory retirement and universities in three chapters of this volume. Note that the focus of judicial rulings on mandatory retirement, particularly by the Supreme Court of Canada, on universities and other public-sector employers is explained by the fact that the Charter guarantees of equality apply only to the actions of governments; regulation of private sector practices falls under the jurisdiction of provincial human rights legislation. The majority judgment in *McKinney* was that the actions of a university do not constitute governmental actions and hence do not warrant Charter protection.

48 See the evidence in Ashenfelter and Card, "The Elimination of Mandatory Retirement." MacGregor in Chapter 2 reports that University of Calgary faculty (who do not face mandatory retirement) have an average retirement age two years higher than that of University of Alberta faculty (who do face mandatory retirement).

[49] The legally permissible mandatory retirement age for faculty members in the United States was raised to 70 in 1982. For a brief chronology of these developments, see Ashenfelter and Card, "The Elimination of Mandatory Retirement," 957–9.

[50] In Chapter 2, MacGregor attributes the Manitoba amendments to retribution by a provincial government for the bitter but successful faculty strike of the mid-1990s.

[51] The Crimson Staff, "The Process of Aging," *The Harvard Crimson*, 23 February 2004, editorial. Also see Rebecca D. O'Brien, "Older Faculty Stay on at Harvard," *The Harvard Crimson*, 12 February 2004, for a detailed account of the rewards and challenges of older professors; 7 percent of Harvard's tenured professors are over 70.

[52] Ruth F. Necheles-Jansyn, "Retirement in Academe: Special Case or Social Model?" *Aging and Work*, 6:3 (1983), 175–85.

[53] See the compelling personal account by McKenzie Leiper in Chapter 11.

[54] The age-wage profile for academics has flattened sharply over the past generation. Krashinsky, "The Case for Eliminating Mandatory Retirement," 47, cites a ratio of beginning to terminal salaries of 2.5 for professors at the University of Toronto in 1984; currently that ratio is less than 2.0 (and close to 1.5 for economists). This likely mirrors growing salary compression at US universities, explained in part by the hike in mandatory retirement age to 70 in 1982 and the abolition of mandatory retirement in 1994.

[55] The Crimson Staff, "The Process of Aging."

[56] See the more detailed analysis of this issue by Klassen with Forgione in Chapter 4.

[57] This complementary role was suggested in Kesselman, "Mandatory Retirement and Older Workers," but one could clearly support a ban on mandatory retirement with no changes in public pension policies; see the comments in Munro's Chapter 10.

[58] The characterizations cited in this and the following two paragraphs are drawn from Chapter 6 by Gunderson and Hyatt, who distinguish between supporters of mandatory retirement per se and supporters of the right to bargain collective agreements that include mandatory retirement.

[59] Mandatory retirement supporters might further argue that the practice involves quid pro quo benefits for workers who agree to the practice, unlike discriminatory practices based on race or sex. However, if mandatory retirement is newly instituted in a workplace, the workers who are approaching age 65 face only the cost of forced retirement with little compensating benefit.

[60] The Supreme Court quoted approvingly the passage: "[T]he facts that all of us once were young, and most expect one day to be fairly old, should neutralize whatever suspicion we might otherwise entertain respecting the multitude of laws… that comparatively advantage [non-aged adults over those aged 65 and up]" (*McKinney* at 297). It expressed this view more directly in a majority decision on another case upholding contractual mandatory retirement: "age differs from other grounds of discrimination since everyone of no matter what religion, colour, social origin, nationality or gender becomes older with the passage of time." *Dickason v. University of Alberta,* [1992] 2 S.C.R. 1103, at 1132–3.

[61] In contrast, Gunderson and Hyatt in Chapter 8 argue, "Since the individual ultimately is applying the rule [of mandatory retirement] to themselves and not to others, this should not be regarded as age discrimination." Yet in collective bargaining the majority of voting union members are applying the rule to others without their consent.

[62] Indeed, the leading Canadian labour economics textbook states, "there is virtually a complete absence of research on the impact that banning [mandatory] retirement will have on other aspects of the employment relationship [besides the length of employment]." Benjamin, Gunderson, and Riddell, *Labour Market Economics,* 411.

[63] Pesando, *The Elimination of Mandatory Retirement,* 3. But quotations similar in spirit could be drawn from Lazear, "Why Is There Mandatory Retirement?," Gunderson and Pesando, "Allowing Mandatory Retirement," Benjamin, Gunderson, and Riddell, *Labour Market Economics*, or Gunderson, "Age Discrimination," and "Banning Mandatory Retirement."

[64] For early economic models of discrimination, see Gary S. Becker, *The Economics of Discrimination*, 2nd ed. (Chicago: University of Chicago Press, 1971) and Kenneth J. Arrow, "Models of Job

Discrimination," in Anthony H. Pascal, ed., *Racial Discrimination in Economic Life* (Lexington, MA: Lexington Books, 1972).

65 This passage from Section 1 of the Charter of Rights and Freedoms provides grounds for exceptions to the prohibitions against discrimination on the enumerated grounds, including age, in section 15.

66 In *GVRD Employees' Union v. GVRD*, [2001] BCCA 435, a justice of the BC Court of Appeal writing for the majority stated: "I would urge the Supreme Court of Canada to reconsider this issue. Eleven years have now passed since *McKinney* was decided. The demographics of the workplace have changed considerably, not only with respect to the university community, but also in the workplace at large.... The extent to which mandatory retirement policies impact on other equality rights, and on the mobility of the workforce, have become prominent social issues. The social and legislative facts now available may well cast doubt on the extent to which the courts should defer to legislative decisions made over a decade ago. The issue is certainly one of national importance." (at para. 127)

Chapter 10 Positive and Personal Choices

1 Bill 68: *An Act to Amend the Provisions of Certain Acts Respecting the Age of Retirement*, 4th Session, 37th Legislature, Province of Ontario, 52 Elizabeth II: 2003.

2 Ontario Ministry of Labour, News Releases: *Providing Choice: a Consultation Paper on Ending Mandatory Retirement*, 18 August 2004: *www.gov.on.ca/LAB/english/news/2004/04-92cp.html* (18 August 2004).

3 Indeed, the home page of the Ontario Human Rights Commission still contains this prominently placed statement: "Nobody has a shelf life. The only thing that's out of date is the idea that older people don't deserve the same respect and opportunities as everyone else. Let's stop age discrimination. It's old news." *www.ohrc.on.ca/english/index.shtml*.

4 Known as BFOR: bona fide occupational requirement. The version of the Ontario Human Rights Code that the Supreme Court used in December 1990 was "1981, S.O. 1981." The version currently in

force is given as "R.S.O. 1990, Chapter H.19," amended in 1993, 1994, 1995, 1997, 1999, 2001, and 2002. Section 9(a) is now section 10(1). See Government of Ontario, Public Statutes (English), Human Rights Code (R.S.O. 1990), *http://192.75.156.68/DBLaws/Statutes/English/ 90h19_e.htm.*

5 Government of Canada, Department of Justice, Canadian Charter of Rights and Freedoms: *www.justice.gc.ca/loireg/charte/const_en.html#recours.* The Canadian Charter of Rights and Freedoms constitutes Schedule B of the Constitution Act, 1982 (79), and the Canada Act 1982 (UK) 1982, c. 11, which came into force on 17 April 1982; section 15 did not do so until 1985.

6 Orley Ashenfelter and David Card, "Did the Elimination of Mandatory Retirement Affect Faculty Retirement?" *The American Economic Review*, 92:4 (Sept. 2002), 957. See below, pp. 211-12

7 Ibid., 957–80.

8 *Statutes of Manitoba*, 1974, C.65, s. 3, 4, 5, 6, 7(2). Prof. Ernest Sirluck, who was president of the University of Manitoba from 1970 to 1976, told me personally (November 2004) that, following the passage of the Human Rights Act in June 1974, he agreed to a request from the University of Manitoba's faculty association (which had been certified that April) to abolish mandatory retirement. Unfortunately, this issue is not discussed in his chapter on his years as president of the University of Manitoba, in *First Generation: An Autobiography* (Toronto and Buffalo, 1996), 304–80. However, Professor Donald McCarthy, former Dean of Arts at the University of Manitoba, and retired professors Lawrence Douglas and Edward Boldt have confirmed (by e-mail dated Thursday, 25 Nov. 2004) that "Sirluck did abolish mandatory retirement; and for a good number of years after the administration [of his successor, Ralph Campbell] had little interest in reinstating it.... As time passed, however, and resources became more scarce, the administration revisited the matter and decided it would be in the University's best interest if there were mandatory retirement. This *would get rid of the highest paid faculty members and free up resources* [my italics added] for other urgent problems. Accordingly, the administration raised the matter during contract negotiations with UMFA,

which agreed to reinstate mandatory retirement." This e-mail text cogently explains why most university administrators want to retain or impose mandatory retirement.

9 Thomas Flanagan, "Policy-Making by Exegesis: the Abolition of 'Mandatory Retirement' in Manitoba," *Canadian Public Policy/ Analyse de Politiques*, 11:1 (1985), 40–53, in particular p. 45, with the quotation cited from *Imogene McIntire v. The University of Manitoba, et al.*, 2 C.H.R.R. (1981) D/305, AT D/309.

10 Flanagan, "Policy-Making," 45–8; he also notes that the "University of Manitoba Faculty Association, after sitting on the sidelines in *McIntire*, did recommend abolition to the Rothstein Commission, but only for a five-year period." (48). His paper was republished in Frederick L. Morton, ed., *Law, Politics, and the Judicial Process in Canada* (University of Calgary Press, 2002). Flanagan was evidently unaware of President Sirluck's abolition of mandatory retirement at the University of Manitoba in 1974.

11 See Jonathan Kesselman, "Time to Retire Mandatory Retirement," Department of Economics Working Paper, University of BC: 26 November 2003, 21, n. 59, citing section 61.1 of the University of Manitoba Act (1996 Amendment). The current contract with the University of Manitoba requires retirement at 69. See also David MacGregor, Chapter 2, and Kesselman, Chapter 9 (which omits some details given in the working paper version).

12 For this and the following, see MacGregor, Chapter 2 and OCUFA (Ontario Confederation of University Faculty Associations) "Mandatory Retirement Discussion Paper" (August 2002), *www.ocufa.on.ca/retirement/retire.asp*. Flanagan, however, cites a different source for the Quebec legislation: Legislature of Quebec, 1982, C.12: *loi sur l'abolition de la retraite obligatoire*.

13 *Dickason v. University of Alberta,* [1992] 2 S.C.R. 1103, whose text is reproduced in: *www.lexum.umontreal.ca/csc-scc/en/pub/1992/vol2/html/ 1992scr2_1103.html* (by Lexum, Université de Montréal). This case is discussed in greater detail in Chapter 3 by Gillin and Klassen and in Chapter 2 by MacGregor.

14 In *McKinney v. University of Guelph,* [1990] 3 S.C.R 229, reproduced in three official Web documents: *www.canlii.org/ca/cas/scc/1990/*

1990scc121.html; *www.canlii.org/ ca/ cas/ scc/ 1990/ 1990scc121.html*, and *www.lexum. umontreal.ca / csc-scc/ en/ pub/ 1990/ vol3/ html/ 1990scr3_0229.html*, from the University of Montreal, which provides the original pagination from the Supreme Court publication. In this latter version, the quotation is from [1990] 3 S.C.R. 229. For references to "bona fide occupational requirements" in this decision, see 3 S.C.R. 229, 291, 308, 412, 439–41.

15 *McKinney*, 232–3.

16 *McKinney*, 275: Justice La Forest. Section 32(1)(b) of the Canadian Charter of Rights and Freedom states that: "This Charter applies ... to the legislature and government of each province in respect of all matters within the authority of the legislature of each province": *www.justice.gc/ loire/ charte/ const_en.html.*

17 See *McKinney*, 258.

18 Jonathan Kesselman, "Mandatory Retirement and Older Workers: Encouraging Longer Working Lives," *C.D. Howe Institute Commentary*, 200 (June 2004), 6; and Kesselman, Chapter 9.

19 *McKinney*, 287–88; partly cited also in Kesselman, "Mandatory Retirement and Older Workers," 6, n. 12: 76 *Dominion Law Reports* [DLR](4th) 545, at 658, 653; and in Kesselman, Chapter 9.

20 Kesselman, "Time to Retire," pp. 6-8 (for quotations) and Chapter 9. He cites the following statement in James Pesando, *The Elimination of Mandatory Retirement: An Economic Perspective* (Toronto: Ontario Economic Council, 1979), p. 23: "[T]he argument that ending compulsory retirement would reduce the job opportunities available in the labour force is not substantiated by economic analysis." Pesando, however, was one of the authorities whom La Forest cited to sustain the majority decision, in *McKinney v. University of Guelph,* in upholding the validity of mandatory retirement.

21 Cited in Klassen and Forgione, Chapter 4. For the Canadian Labour Congress resolution of 1980, stating its opposition to any attempts to eliminate mandatory retirement, see also below, pp. 16.

22 CANSIM II @ Chass: *www.chass.utoronto.ca/*.

23 *McKinney*, 242, 402. On the relationship between government financing of universities and mandatory retirement, see below pp. 212-15.

24 Howard Fink, "Québec Universities Without Mandatory Retirement:

Policies and Experiences," in Peter Russell and Ken Rea, eds., *Redesigning Retirement: Proceedings of a Joint Forum Presented by the University of Toronto Faculty Association and the Retired Academics and Libarians at the University of Toronto: Innis Town Hall, Saturday 5 April 2002* (Toronto, 2003), 11–13. Online version: *www.ralut.ca/proc.pdf.*

25 Kesselman, Chapter 9, citing Frank Reid, "Economic Aspects of Mandatory Retirement: the Canadian Experience," *Relations industrielles*, 43:1 (1988), 101–13.

26 M. Shannon and D. Grierson, "Mandatory Retirement and Older Worker Employment," *Canadian Journal of Economics*, 37:3 (August 2004), 528–51 (quotation on 550).

27 Ashenfelter and Card, "Mandatory Retirement," 967–9, especially Table 2, 968.

28 In her dissent, Justice Claire L'Heureux-Dubé observed that "since the number of people who attain that age [of 65], and wish to continue working after that age and [are] physically and intellectually capable of doing so, is not overwhelming, it is difficult to conclude that the labour force will be adversely affected." *McKinney*, 235, 435.

29 Kesselman, "Time to Retire," 8–11, and especially n. 18. See also Kesselman, Chapter 9.

30 *McKinney*, 289–97.

31 *McKinney*, 289, 309. The specific work cited is Morley Gunderson and James Pesando, eds., *Eliminating Mandatory Retirement: Economic and Human Rights* (Toronto: Faculty of Management Studies, University of Toronto, 1980). See also Morley Gunderson and James Pesando, "The Case for Allowing Mandatory Retirement," *Canadian Public Policy/Analyse de politiques*, 14:1 (March 1988), 32–9.

32 Kesselman, "Time to Retire," 8 (for quotation); and Kesselman, Chapter 9.

33 *McKinney,* 427–8.

34 *McKinney*, 393.

35 Quotations from Kesselman, "Time to Retire", 9; see also Kesselman, Chapter 9 (n. 33), citing Josef Richter, "Economic Aspects of Ageing: A Review of the Literature," in Georg Stolnitz, ed., *Demographic Causes and Economic Consequences of Population Aging: Europe and North America* (New York: United Nations, 1992), 181.

[36] Kesselman, Chapter 9 (n. 35) cites a 1995 Labour Canada report, which states that "poor performers are usually unhappy in their jobs and are anxious to leave as soon as feasible."

[37] In my own case, with a university career so extending from 1964 (age 26) to 2005 (age 67, i.e., post formal retirement), 29 of my 81 publications of journal articles and essays (excluding book reviews, and earlier monographs, etc.), or 35.8 per cent, have appeared (or will appear in 2005) since I turned 60, in 1998.

[38] Ashenfelter and Card, "Elimination of Mandatory Retirement," 976–7. The Ashenfelter-Card study was based exclusively on American universities using TIAA-CREF, the country's largest defined contribution pension system (one in which the employees own and control the pension fund, as opposed to a defined benefit scheme controlled by the employers).

[39] The *Crimson* editorial (*The Process of Aging*) can still be found online at: *www.thecrimson.com/article.aspx?ref=357661.*

[40] *McKinney v. University of Guelph,* [1990] 3 S.C.R. 229, 235; and 425.

[41] David Foot: Quoted in Rosanna Tamburri, "Rethinking the Rules on Retirement," in *University Affairs,* December 2003, 13.

[42] *McKinney*, 430–1, cited in part in Kesselman, "Time to Retire," p. 20; see also Kesselman, Chapter 9.

[43] *McKinney*, 427–8; and also her comments on p. 435. In 1997–98 (latest available data), the average salary for a full professor at the University of Toronto ($102,800 CAD), generally regarded as Canada's leading university, was only 77 percent of the mean of average salaries for full professors in ten comparable public universities in the U.S. ($133,220 CAD). University of Toronto Faculty Association, *How Competitive Are Our Salaries?*, UTFA News Bulletin (9 April 1999), *www.utfa.org/ html/ newsbul/ html/ apr0999.htm.*

[44] See Martin Friedland, *University of Toronto: A History* (Toronto, 2002), 565. That, then, meant the age of 68.

[45] At the University of Toronto, student evaluations of teaching were first undertaken by the Department of Political Economy in 1965–66, and then by other departments in the Faculty of Arts and Science in 1966–67, under the administration of the Students Administrative Council (SAC). Merit award increases, combined

with Across the Board increases (COLA), began in 1972. See Friedland, *History*, 531; William H. Nelson, *The Search for Faculty Power: the History of the University of Toronto Faculty Association, 1942–1992* (Toronto, UTFA: 1993), 78–81.

[46] In *McKinney* 426–7. She also commented that: "the value of tenure is threatened by incompetence, not by the aging process."

[47] See n. 36 above.

[48] See Gunderson and Pesando, "Allowing Mandatory Retirement," 32–9; and other publications cited in n. 31 above; Morley Gunderson, *Flexible Retirement as an Alternative to 65 and Out*, C.D. Howe Institute Commentary 106 (Toronto, 1998); Morley Gunderson, "Age Discrimination in Employment in Canada," *Contemporary Economic Policy*, 21:3 (July 2003), 318–28; Morley Gunderson, "Banning Mandatory Retirement: Throwing Out the Baby with the Bathwater," *C.D. Howe Institute Backgrounder* 79 (Toronto: C.D. Howe Institute, 2004). See also Gunderson and Hyatt, Chapter 8.

[49] *McKinney,* 233, 244, 266, 269, 272, 379, 388 and *Dickason,* 1132, 1167–69, 1173–74.

[50] Friedland, *History,* 234. To make his subsequent point, Friedland cites (on p. 125) the case of the philosophy professor, James Hume, "considered a disaster," who, after 37 years, "was forced to retire at age 65 [in 1926], when almost everyone else in his position was granted an extension."

[51] Nelson, *Faculty Power*, 155, 15, respectively.

[52] Friedland, *History*, 543–54.

[53] Nelson, *Faculty Power*, 155. For the Memorandum of Agreement, see 93–112; and for the de facto binding arbitration achieved in 1982, see 113–34.

[54] Friedland, *History,* 563–7, 584.

[55] Nelson, *Faculty Power*, 155. The current (revised November 2002) CAUT resolution states, "Mandatory retirement is discrimination on the basis of age, and may give rise to discrimination on the basis of sex or other grounds. Academic staff have a right to continue their employment beyond the standard retirement age under the same terms and conditions." See the online document at: *www.caut.ca/english/about/policy/retirement.asp.*

[56] The Memorandum of Agreement may be found as a document on the Web site of UTFA: *www.utfa.org/*. This agreement speciously suggests that, with permission of the chair and dean, a faculty member may continue with his/her employment until age 68—though only on condition that the dean and chair find and provide the necessary funding, since the professor's salary is removed from the departmental budget on retirement. Needless to say, very, very few professors have been able to enjoy this privilege, chiefly those who bring research funds to the university.

[57] Nelson, *Faculty Power*, 154.

[58] See *McKinney*, 230.

[59] Cited by Justice La Forest in *McKinney*, 313. For a further discussion of this issue, see Klassen and Forgione, Chapter 4; and below, pp. 215-217.

[60] *McKinney*, 406.

[61] Gunderson and Hyatt, Chapter 8.

[62] Justice La Forest, in *McKinney* at p. 292: "[T]here has long been a differentiation made between it and other rights, and that like other rights, it is not absolute. Under the Charter, however, questions as to whether these qualifications have been made must be measured against the requirements of s. 1 of that instrument"; and at p. 297: [T]here is nothing inherent in most of the specified grounds of discrimination, e.g., race, colour, religion, national or ethnic origin, or sex that supports any general correlation between those characteristics and ability. But that is not the case with age. There is a general relationship between advancing age and declining ability," an argument whose validity was disputed above. See the more detailed analysis of this issue in Gillin and Klassen, Chapter 3.

[63] George Orwell, *Animal Farm: A Fairy Story* (London: Secker & Warburg, 1945; reprinted 1961), 105. This new and now the farm's single commandment, replacing all others, was stated in capital letters.

[64] Kesselman, "Time to Retire," 3–4 (for quotations); and Kesselman, Chapter 9, in particular citing Krashinsky, "The Case for Eliminating Mandatory Retirement," Why Economics and Human Rights Need Not Conflict," *Canadian Public Policy/Analyse de politiques*, 14:1 (March 1988), 40–51, on the inability of employees

properly to predict their future circumstances. See nn. 49-49 above.

65 See the remarks of Prof. Angela Hildyard, Vice-President Human Resources, at the UTFA-RALUT conference of 5 April 2003, in Russell and Rea, *Redesigning Retirement*, 14–15: "Then finally one of the concerns that I have is equity and diversity. We do rely on retirement within all of our staff groups but particularly within the faculty as a way for us to start to increase the diversity of the faculty on this campus. The diversity of our students is huge. Our faculty diversity does not match our student population and we do rely on the[se] retirements to[o] in an attempt to bring more diverse faculty on this campus."

66 See above, n. 6 and related text.

67 See also Meyer Brownstone, "Forum: Time to Go—Mandatory Retirement, Who Decides?" *The Bulletin of the University of Toronto*, 57:16 (22 March 2004), 16, in which he asks "[W]hat is the basis of excluding age as a highly significant element in diversity?"

68 Government of Canada, Department of Justice, Canadian Charter of Rights and Freedoms: *www.justice.gc.ca/loireg/charte/const_en.html#recours.*

69 *McKinney*, 403.

70 See above, n. 6 and related text; and also *McKinney*, 427, referring to 29 U.S.C. \SS\ 631(d).

71 See Kesselman, Chapter 9, n. 54, citing Michael Krashinksy, "The Case for Eliminating Mandatory Retirement," 40–51, to the effect that, in 1984, the ratio of final salaries for 65-year-old professors was then 2.5 times the average salary for newly hired assistant professors. See also Nelson, *Faculty Power*, 79–82, for the introduction of PTR (merit award increases) at the University of Toronto, in 1972.

72 See Friedland, *History*, 560–61, 581–2; See also Nelson, *Faculty Power*, 124–5.

73 See Kesselman, Chapter 9.

74 See n. 37 above.

75 James Turk: Quoted in Tamburri, "Rethinking the Rules on Retirement," 13.

76 *McKinney*, 437.

[77] John Myles, "Changing Public Policies, Changing Market Policies," in Peter Russell and Ken Rea, eds., *Redesigning Retirement* (Toronto: RALUT, 2003), 4–6, *www.ralut.ca/proc.pdf.*

[78] See above, p. 210 and n. 59.

[79] Kesselman, "Time to Retire," 14–17; but, in responding to my objections in this chapter, Kesselman, in Chapter 9, adopts a much more nuanced position, which does provide a good case for contending that society would enjoy significant economic benefits, if more of the educated workforce chose—voluntarily—to continue working past normal retirement age.

[80] As noted earlier, the current average age of retirement, estimated for all Quebec universities, after twenty years without contractual mandatory retirement, is 63.5. See n. 20 above.

Chapter 11 "A Frivolous and Vexatious Complaint?"

[1] On 26 March 2004, King's College was renamed King's University College at the University of Western Ontario.

[2] Campbell Clark, "Martin pledges new priorities," *The Globe and Mail*, 13 December 2003, A1. Also, Canada's head of state, Her Majesty Queen Elizabeth II, celebrated her 78th birthday on April 21, 2004 and appears headed for many more years as reigning monarch.

[3] Erin Anderssen, "Nation Builders of 2003," *The Globe and Mail*, 13 December 2003, F1.

[4] Rick Groen. "Close up," *The Globe and Mail*, 13 December 2003, R1.

[5] Campbell Clark and Drew Fagan, "PM Hopes to Rule at Age 75," *The Globe and Mail*, 19 December 2003, A1; Steven Chase, "Martin Opposes Mandatory Retirement," *The Globe and Mail*, 20 December 2003, A2.

[6] Kim Lunman, "Broadbent's Bid for Comeback Pumps up NDP," *The Globe and Mail*, 19 December 2003, A7.

[7] Prime Minister Chrétien was an exception to this statement during his last few years in office. When he reached his late sixties, Paul Martin and others publicized his age in an effort to push him out of office. In an interesting paradox, Martin now promotes himself as an elder statesman ready to rule until age seventy-five. Perhaps the magic age of sixty-five has become a double-edged sword.

[8] Several observers have noted that the mean age of the Supreme Court justices dealing with the *McKinney* case was sixty-five.

[9] Political leaders are not the only people who enjoy the opportunity to remain productive beyond a designated retirement age. Many writers, media personalities, and political appointees continue to work well beyond age sixty-five. For example, in 2003, P.D. James published a first-rate novel at age eighty-three, and, in March of the following year, Alistair Cooke retired three weeks before his death at age 95. Alan Greenspan, as chair of the US Federal Reserve Board, holds enormous power over the world economy; he was first appointed at age sixty-two, and has recently been confirmed for a fifth term, at age seventy-eight

[10] Many women develop career paths that depart from the widely accepted linear model. For more information about these alternative careers see Judi Marshall, "Re-Visioning Career Concepts: A Feminist Invitation," in *Handbook of Career Theory,* ed. Michael B. Arthur, D.T. Hall, and B.S. Lawrence (Cambridge: Cambridge University Press, 1989) and Michael B. Arthur, Kerr Inkson, and Judith K. Pringle, *The New Careers: Individual Action and Economic Change* (London: Sage, 1996).

[11] During my final year, major Canadian companies visited the university to recruit graduates for their management training courses, but it was understood that they were interested only in male graduates. Teaching was one of the few employment options for women graduates at that time.

[12] These norms were perpetuated in a variety of ways. For example, in the early 1960s, the University Women's Club in my community held a heated debate on the topic, "Should women 'work'?" Members of the club generally agreed that women should stay home with their children and avoid taking on the real work of the world in the labour force.

[13] I now appreciate the impact of Jonathan Kesselman's observation that mandatory retirement clears the way for younger workers in the same way that the earlier male breadwinner model kept many women out of the labour force. I have experienced both forms of discrimination. See Chapter 9.

[14] I had not taken sociology courses during my undergraduate years but slowly began to realize that the discipline held a great deal of appeal. In 1964 I read Betty Friedan's groundbreaking work, *The Feminine Mystique*, (New York: Dell, 1963), and began to question many of the prevailing expectations of women's roles. I still have that dog-eared, yellowed copy of her book as a reminder of the impact of her words.

[15] During my child-rearing years, I had become very involved in community affairs, sitting on various boards and serving several terms as an elected municipal councillor.

[16] As an older graduate I was not naïve about my prospects for employment. I had purposely taken advanced statistics and research methods courses, and chosen a PhD supervisor well-schooled in these areas, because I knew most sociology departments had difficulty recruiting faculty with strong quantitative backgrounds.

[17] I should stress that these observations do not constitute an attempt to assign blame. The chair of our department was aware of my discomfiture with the process and, in fact, shared my dread of retirement, but his only option at the time was to fall in line with the institution's long-range plans.

[18] This restriction has generated a great deal of debate, and the Chief Commissioner of the Ontario Human Rights Commission, Keith Norton, has lobbied for change for many years.

[19] *McKinney vs The University of Guelph*, 1990 remains the landmark Supreme Court case dealing with mandatory retirement for university professors. For more detail, see Gillin and Klassen (Chapter 3).

[20] See Virginia Galt, "Retirees Increasingly Are Keen to Return to Work," *The Globe and Mail*, 30 January 2004, B1. This report also triggered an article in the University of Western Ontario's newspaper (Karmen Dowling, "Professor Challenges Mandatory Retirement," *Western News*, 40:5, 5 February 2004), and a second follow-up article appeared in Western's alumni publication ("Professor Challenges Mandatory Retirement," *Western Alumni Gazette*, Spring 2004, 5).

[21] On 20 March 2004, the lead editorial in the *The Globe and Mail* summarized the arguments for ending the practice, citing as

justification a shrinking supply of younger workers outflanked by a growing number of healthy sixty-five-year-olds with the potential to lead long, productive lives. The editor emphasized the necessity to grant older workers the same rights and freedoms as other previously disadvantaged groups:

> Canada is becoming more and more diverse. Soon, men may be able to marry men and women, women. New immigrants are changing what Canada looks like. There is a confidence in this country that, though all this change brings uncertainty, it also promotes individuality and creativity, and is creating a more vigorous, stimulating place to live. It's only fair to apply that same kind of thinking to older workers.

("A few reasons to end mandatory retirement," *The Globe and Mail*, 20 March 2004, A24.)

22 I used the federal research grant to conduct interviews with women practising law across the province. Ironically, this research focused in part on the unique career paths followed by many women. Their experiences provided the material for many of the academic papers, and my book was the culmination of ten years of research.

Chapter 12 Conclusion

1 Government of Ontario, Ministry of Labour, "Providing Choice: A Consultation Process on Ending Mandatory Retirement," 1.

2 Peter H. Russell, "High Courts and the Rights of Aboriginal Peoples: The Limits of Judicial Independence," 61 *Saskatchewan Law Review* (1998), 268.

3 *Canadian Charter of Rights and Freedoms*, section 15. The other enumerated grounds of unconstitutional discrimination are race, national or ethnic origin, colour, religion, sex, and mental or physical disability.

4 *The Queen v. Oakes,* (1986) 1 S.C.R. 103.

Selected Bibliography

This bibliography represents the first attempt to compile primarily Canadian Writings on mandatory retirement.

Adams, George W. *Mandatory Retirement and Constitutional Choices.* Kingston, Ontario: Industrial Relations Centre, Queen's University, 1992.

Anderson, Gavin W. "Filling the 'Charter Gap' Human Rights Codes in the private sector." *Osgoode Hall Law Journal* 33:4 (1995), 749–83.

Ashenfelter, Orley and David Card. "Did the Elimination of Mandatory Retirement Affect Faculty Retirement?" *The American Economic Review* 92:4 (2002), 957–80.

Atchenson, M. Elizabeth and Lynne Sullivan. "Passage to Retirement: Age Discrimination and the Charter." In Anne Bayefsky and Mary Eberts, eds., *Equality Rights and the Canadian Charter of Rights and Freedoms.* Toronto: Carswell, 1985, 231–92.

Baum, David. *The Final Plateau: The Betrayal of Our Older Citizens.* Toronto: Burns and MacEachern, 1974.

Beatty, David. "The Canadian Conception of Discrimination." *Canadian Labour & Employment Law Journal* 4 (1996), 263–82.

Beaujot, Roderic. "Projecting the Future of Canada's Population: Assumptions, Implications and Policy." *Canadian Studies in Population* 30 (2003), 1–28.

Beaujot, Roderic, Suzanne Shiel, and Lorraine Schoel. *Immigration and the Population of Canada.* Report Prepared for Immigration Policy Branch, Employment and Immigration. Ottawa: Employment and Immigration Canada, 1989.

Beaujot, Roderic and Feng Hou. *Projecting the Visible Minority Population of Canada: The Immigration Component.* Ottawa: Statistics Canada, 1993.

Bélanger, Alain. *Report on the Demographic Situation in Canada 2000.* Cat. No. 91-209. Ottawa: Statistics Canada, 2001.

Bellamy, Edward. *Looking Backward.* New York: Dover Publications, 1996 (orig. pub. 1888).

Bers, Melvin K. *Union Policy and the Older Worker.* Berkeley: Institute of Industrial Relations, University of California, Berkeley, 1957.

Blackburn, Robert T. and Janet H. Lawrence. *Faculty at Work: Motivation, Expectation, Satisfaction.* Baltimore: Johns Hopkins University Press, 1995.

Blane, David *et al.* "Inequalities in Quality of Life in Early Old Age." *Research Findings: 9.* ESRC Growing Older Programme. Department of Sociological Studies, University of Sheffield, UK, 2002.

Blau, David. "Labor Force Dynamics of Older Men." *Econometrica* 62 (1994), 117–56.

Blondal, Sveinbjorn and Stefano Scarpetta. *The Retirement Decision in OECD Countries.* Economics Department Working Paper Series, no. 202. Paris: Organization for Economic Cooperation and Development, 1998.

Bradford, Neil. "The Policy Influence of Economic Ideas: Interests, Institutions and Innovation in Canada" *Studies in Political Economy* 59 (Summer 1999), 17–59.

Briskin, Linda and Patricia McDermott, eds. *Women Challenging Unions: Feminism, Democracy and Militancy.* Toronto: University of Toronto Press, 1993.

Brown, Richard. "Human Rights in Employment: Of Participation and Compensation." *Canadian Labour & Employment Law Journal* 4 (1996), 283–309.

Brownstone, Meyer. "Forum: Time to Go—Mandatory Retirement, Who Decides?" *The Bulletin of the University of Toronto* 57:16 (22 March 2004), 16.

Butler, Robert N. "Ageism: Another Form of Bigotry." *The Gerontologist* 9:3 (1969), 243–6.

Canadian Human Rights Review Panel. *Promoting Equity: A New Vision*. Ottawa: Canadian Human Rights Review Panel under the authority of the Minister of Justice and the Attorney General of Canada, 23 June 2000.

Carmichael, Lorne. "Firm–Specific Human Capital and Promotion Ladders." *Bell Journal of Economics* 14 (1983), 251–61.

Carroll, William K. *Corporate Power in a Globalizing World*. Toronto: Oxford University Press, 2004.

Charles, Enid, Nathan Keyfitz, and H. Roseborough. *The Future Population of Canada*. Ottawa: Dominion Bureau of Statistics, Bulletin No. FB4, 1946.

Cheal, David, ed. *Aging and Demographic Change in Canadian Context*. Toronto: University of Toronto Press, 2003.

Chen, Jiajian and Wayne J. Millar. "Are Recent Cohorts Healthier Than Their Predecessors?" *Health Reports* 11 (2000), 9–23.

Chesnais, Jean-Claude. *L'inversion de la pyramide des âges en Europe: perspectives et problèmes*. In International Population Conference (vol 3). Liège, Belgium: IUSSP, 1989, 53–68.

Choi, Namkee G. "Self–Defined Retirement Status and Engagement in Paid Work Among Older Working-Age Women: Comparison between Childless Women and Mothers." *Sociological Inquiry* 72:1 (Winter 2002), 43–71.

Chotalia, Shirish P. "The Supreme Court and Mandatory Retirement: Sanctioning the Status Quo." *Constitutional Forum* 4:3 (1993), 67–70.

Citizenship and Immigration. *Pursuing Canada's Commitment to Immigration: The Immigration Plan for 2002*. Ottawa: Citizenship and Immigration Canada, 2001.

Connidis, Ingrid and Julie McMullin. "Sociological Ambivalence and Family Ties: A Critical Perspective." *Journal of Marriage and the Family* 64:3 (August 2002), 558–67.

Copperman, Lois F. and Frederick D. Keast. *Adjusting to an Older Work Force*. New York: Van Nostrand (1983), 150–51.

Cormier, Jeffrey. *The Canadianization Movement: Emergence, Survival, and Success*. Toronto: University of Toronto Press, 2004.

Crane, M.C. "Human Rights, *Bona Fide* Occupational Requirements and the Duty to Accommodate: Semantics or Substance?" *Canadian Labour & Employment Law Journal* 4 (1996), 209–232.

Creese, Gillian. "Gendering Collective Bargaining: From Men's Rights to Women's Issues." *The Canadian Review of Sociology and Anthropology* 33 (1996), 437–56.

Cranswick, Kelly. *Aging and Social Support: General Social Survey.* Cat. No. 89–583–XIE. Ottawa: Statistics Canada, 2003.

Croll, David. *Retirement Without Tears: Report of the Special Senate Committee on Retirement Age Policies.* Ottawa: Supply and Services, 1979.

Cruikshank, Margaret. *Learning to be Old: Gender, Culture and Aging.* Lanham, MD: Rowman & Littlefield Publishers, 2003.

Davis, Kingsley and P. van den Oever. "Age Relations and Public Policy in Advanced Industrial Societies." *Population and Development Review* 7 (1981), 1–18.

de Beauvoir, Simone. *The Coming of Age.* New York: G.P. Putnam's Sons, 1972.

Dentinger, Emma and Marin Clarkberg. "Informal Caregiving and Retirement Timing Among Men and Women." *Journal of Family Issues* 23:7 (2002), 857–79.

Denton, Frank T., Christine H. Feaver, and Byron Spencer. "The Future Population of Canada, its Age Distribution and Dependency Relations." *Canadian Journal on Aging* 17 (1998), 83–109.

Denton, Frank, and Byron G. Spencer. "Population Aging and its Economic Costs: A Survey of the Issues and Evidence." *Canadian Journal on Aging* 19 (Supplement Summer 2000), 1–31.

Denton, Frank T. and Byron G. Spencer. "Some Demographic Consequences of Revising the Definition of 'Old Age' to Reflect Future Changes in Life Table Probabilities." *Canadian Journal on Aging* 21 (2002), 349–56.

Department of Health and Welfare. *Early Retirement: A Preliminary Analysis.* Ottawa: Health and Welfare, 1993.

Diamond, Peter. "A Framework for Social Security Analysis." *Journal of Public Economics* 8: 3 (1977), 275–98.

Diamond, Peter, and Botond Köszegi. "Quasi–Hyperbolic

Discounting and Retirement." *Journal of Public Economics* 87: 9–10 (2003), 1839–72.

Dorfman, Lorraine T. "Stayers and Leavers: Professors in an Era of No Mandatory Retirement." *Educational Gerontology* 28:1 (2002) 15–33.

Drummond, Robert J. "Comment on 'Mandatory Retirement: Intergenerational Justice and the Canadian Charter of Rights and Freedoms' by Samuel LaSelva." *Canadian Journal of Political Science* 21:3 (1988), 583–95.

Duchesne, Doreen. "More Seniors at Work." *Perspectives on Labour and Income*. Statistics Canada, Cat. No. 75–001–XIE, 5, 2 (February 2004), 3–17.

Duffy, Ann and Norene Pupo. *Part-time Paradox: Connecting Gender, Work and Family*. Toronto: McClelland and Stewart, 1992.

Duffy, Ann, Daniel Glenday and Norene Pupo. "Seniors in the Part–Time Labour Force: Issues of Choice and Power." *International Journal of Canadian Studies* 18 (Fall 1998), 67–86.

Dunlop, P. *Mandatory Retirement Policy: A Human Rights Dilemma?* Ottawa: Conference Board of Canada, 1980.

Economic Council of Canada. *One in Three: Pensions for Canadians to 2030*. Ottawa: Ministry of Supply and Services, 1979.

Fast, Janet and Judith Frederick. *The Time of Our Lives: Juggling Work and Leisure Over the Life Cycle*. Ottawa: Statistics Canada, Cat. No. 89–584–MIE, Number 4, 1998.

Feldman, Daniel C. and Kim Seongsu. "Bridge Employment During Retirement: A Field Study of Individual and Organizational Experiences with Post–Retirement Employment." *Human Resource Planning* 23:1 (2000), 14–26.

Fink, Howard. "Quebec Universities Without Mandatory Retirement: Policies and Experiences," in Peter Russell and Ken Rea, eds., *Redesigning Retirement: Proceedings of a Joint Forum Presented by the University of Toronto Faculty Association and the Retired Academics and Librarians at the University of Toronto: Innis Town Hall, Saturday 5 April 2003* (Toronto: RALUT, 2003), pp. 11–13.

Finkelstein, Neil and Geoffrey Howard. "Retirement as a Case Study." *The Advocates' Quarterly* 9:2 (1998), 142–59.

Flanagan, Thomas. "Policy-Making by Exegesis: The Abolition of 'Mandatory Retirement' in Manitoba." *Canadian Public Policy* 11:1 (1985), 40–53.

Foucault, Michel. *The Order of Things: An Archaeology of the Human Sciences.* New York: Vintage, 1973.

Fourzly, Michel and Marc Gervais. *Collective Agreements and Older Workers in Canada.* Ottawa: Labour Program, Human Resources Development Canada, 2002.

Friedan, Betty. *The Fountain of Age.* New York: Simon and Schuster, 1993.

Friedland, Martin. *University of Toronto: A History.* Toronto: University of Toronto Press, 2002.

Frost, Ann. "Union Involvement in Workplace Decision Making: Implications for Union Democracy." *Journal of Labor Research* 21.2 (2000), 167–187.

Gee, Ellen M. and Gloria Gutman, eds. *The Overselling of Population Aging: Apocalyptic Demography, Intergenerational Challenges and Social Policy.* Don Mills, Ontario: Oxford University Press, 2000.

Gillin, C.T. and Thomas R. Klassen. "Age Discrimination and Mandatory Retirement Policies: A Comparison of Labor Market Regulation in Canada and the United States." *Journal of Aging and Social Policy* 7, 1 (1995), 85–102.

Gillin, C.T. and Thomas R. Klassen. "Retire Mandatory Retirement." *Policy Options* 21, 6 (2000), 59–62.

Gomez, Rafael, Morley Gunderson, and Andrew Luchak. "Mandatory Retirement: A Constraint in Transitions to Retirement?" *Employee Relations* 24 (2002), 403–422.

Gower, Dave. "Income Transition Upon Retirement." *Perspectives on Labour and Income.* Statistics Canada, Cat. No. 75–001–XPE, 10, 4 (Winter 1998), 18–23.

Graebner, William. "Retirement and the Origins of Age Discrimination." In Meredith Minkler and Carroll L. Estes, eds., *Readings in the Political Economy of Aging.* Farmingdale, New York: Baywood Publishing Company, 1984, 177–96.

Graebner, William. *A History of Retirement: The Meaning and Function of an American Institution, 1885–1978.* New Haven, Conn.: Yale University Press, 1980.

Grant, Hugh M. with Gretta Wong Grant. *Age Discrimination and the Employment Rights of Elderly Canadian Immigrants.* A report for the Law Commission of Canada, 2002.

Greene, Ian. *The Charter of Rights.* Toronto: James Lorimer & Company Ltd., Publishers, 1989.

Guest, Denis. *The Emergence of Social Security in Canada*, 3rd ed. Vancouver: University of British Columbia Press, 2003.

Guillemette, Yvan. *Follow Quebec's Lead: Removing Disincentives to Work after 60 by Reforming the CPP/QPP.* Commentary #199. Toronto: C.D. Howe Institute, 2004.

Gunderson, Morley. "Age Discrimination in Employment in Canada". *Contemporary Economic Policy* 21:3 (2003), 318–28.

Gunderson, Morley. "Banning Mandatory Retirement: Throwing Out the Baby with the Bathwater." *C.D. Howe Institute Backgrounder* 79. Toronto: C.D. Howe Institute, 2004.

Gunderson, Morley. *Flexible Retirement as an Alternative to 65 and Out.* Toronto: C.D. Howe Institute, 1998.

Gunderson, Morley. "Mandatory Retirement and Personnel Policies." *Columbia Journal of World Business* 28:2 (1983), 8–15.

Gunderson, Morley and James Pesando. "Eliminating Mandatory Retirement: Economics and Human Rights." *Canadian Public Policy* 6:2 (1980), 352–60.

Gunderson, Morley and James Pesando. "The Case for Allowing Mandatory Retirement." *Canadian Public Policy* 14:1 (1988), 32–9.

Guppy, Neil. "The Magic of 65: Issue and Evidence in the Mandatory Retirement Debate." *Canadian Journal of Aging* 8 (1989), 173–86.

Habtu, Roman. "Men 55 and Older: Work or Retire?" *Perspectives on Labour and Income* 15:1 (2003), 47–54.

Hall, G. Stanley. *Senescence: The Last Half of Life.* New York: D. Appleton and Company, 1922.

Hatch, Laurie Russell. *Beyond Gender Differences: Adaptation to Aging in Life Course Perspective.* Amityville, New York: Baywood Publishing Company, 2000.

Hayes, Bernadette C. and Audrey Vanderheuval. "Attitudes Toward Mandatory Retirement: An International Comparison." *International Journal of Aging and Human Development* 39 (1994), 209–31.

He, Y. Helen, Angela Colantonio and A. Victor W. Marshall. "Later–Life Career Disruption and Self–Related Health: An Analysis of General Social Survey Data." *Canadian Journal on Aging* 22:1 (2002), 45–58.

Health and Welfare Canada. *Review of Demography and its Implications for Economic and Social Policy. Charting Canada's Future.* Ottawa: 1989.

Hewitt Associates, *Mandatory Retirement: Current Practice and Future Directions Survey Report.* Toronto 2003.

Hicks, Peter. "New Policy Research on Population Aging and Life-Course Flexibility." *Horizons* 6 (2003), 3–6.

Hiebert, Janet L. *Charter Conflicts: What is Parliament's Role?* Montreal: McGill–Queen's University Press, 2002.

Hogg, David. *Constitutional Law in Canada*, 4th ed. Toronto: Carswell, 1997.

Horkeimer, Max and Theodor Adorno. *The Dialectic of Enlightenment.* London: New Left Books, 1972 (orig. pub. 1947).

Human Resources Development Canada. *Challenges of an Aging Workforce: An Overview of the Issue.* Ottawa: Human Resources Development Canada, 2002.

Hurd, W.B. "Some Implications of Prospective Population Changes in Canada." *Canadian Journal of Economics and Political Science* 5 (1939), 492–4.

Jackson, Andrew, David Robinson, with Bob Baldwin and Cindy Wiggins. *Falling Behind: The State of Working Canada, 2000.* Ottawa: Canadian Centre for Policy Alternatives, 2000.

Kahn, Paul W. *The Cultural Study of Law: Reconstructing Legal Scholarship.* Chicago: The University of Chicago Press, 1999.

Katz, Stephen. *Disciplining Old Age: The Formation of Gerontological Knowledge.* Charlottesville: University Press of Virginia, 1996.

Kausler, Donald H. "Motivation, Human Aging, and Cognitive Performance." In James E. Birren and K. Warner Schaie, eds., *Handbook of the Psychology of Aging*, 3rd ed., San Diego, CA: Academic Press, 1990, 172–83.

Kesselman, Marc L. "Putting the Professor to Bed: Mandatory Retirement of Tenured Faculty in the United States and Canada." *Comparative Labour Law Journal* 17 (1995), 206–49.

Kesselman, Jonathan. *Mandatory Retirement and Older Workers: Encouraging*

Longer Working Lives. Toronto: C.D. Howe Institute, 2004.

Khan, Anwar N. "Mandatory Retirement Age for University Professors." *Journal of Law and Education* 19:1 (1990), 35–146.

Kieran, Patrick."Early Retirement Trends." *Perspectives on Labour and Income* 2:9 (2001), 5–11.

Kite, Mary E. and Lisa Smith Wagner. "Attitudes Toward Older Adults." In Todd D. Nelson, ed., *Ageism*. Cambridge, Mass.: MIT Press, 2002, 129–61.

Klassen, Thomas R. "The End of Mandatory Retirement and Beyond," *Benefits and Pensions Monitor* 14:6 (2004), 22–3.

Klassen, Thomas R. and C.T. Gillin. "The Heavy Hand of the Law: The Canadian Supreme Court and Mandatory Retirement." *Canadian Journal on Aging/La Revue cannadienne du viellissement* 18:2 (1999), 259–76.

Kleemeier, R.W. *Ageing and Leisure*. New York: Oxford University Press, 1961.

Knopff, Rainer and F. L. Morton. *Charter Politics*. Toronto: Nelson Canada, 1992.

Kohli, Martin, Martin Rein, Anne–Marie Guillemand, and Herman van Gunsteren. *Time for Retirement: Comparative Studies of Early Exit from the Labor Force*. Cambridge: Cambridge University Press, 1991.

Krashinsky, Michael. "The Case for Eliminating Mandatory Retirement: Why Economics and Human Rights Need Not Conflict." *Canadian Public Policy* 14:1 (1988), 40–51.

Labour Canada. *An Industrial Relations Perspective on Mandatory Retirement*. Ottawa: Labour Canada, 1985.

LaSelva, Samuel V. "Mandatory Retirement: Intergenerational Justice and the Canadian Charter of Rights and Freedoms." *Canadian Journal of Political Science* 20:1 (1987), 149–62.

Laslett, Peter. *A Fresh Map of Life: The Emergence of the Third Age*. Cambridge, MA: Harvard University Press, 1991.

Law Commission of Canada, "Law and Relationships between Generations." *Discussion Paper*. Ottawa: Law Commission of Canada, February 2004.

Lazear, Edward. "Why is There Mandatory Retirement." *Journal of Political Economy* 87:6 (1979), 1261–84.

LeBlanc, L. Susanne and Julie Ann McMullin. "Falling Through the Cracks: Addressing the Needs of Individuals between Employment and Retirement." *Canadian Public Policy* 23 (1997), 289–304.

Leontief, Wassily. "Theoretical Assumptions and Nonobserved Facts." *American Economic Review* 61:1 (March 1971), 1–7.

Lepofsky, M. David. "The Canadian Judicial Approach to Equality Rights: Freedom Ride or Roller Coaster?" *Law and Contemporary Problems* 55:1 (1992), 167–99.

Levine, Martin Lyon. *Age Discrimination and the Mandatory Retirement Controversy.* Baltimore: Johns Hopkins University Press, 1988.

Lowe, Graham. "Retirement Attitudes, Plans and Behavior." *Perspectives on Labour and Income* 3:3 (1991), 8–17.

Luchak, Andrew. "Retirement Plans and Pensions: An Empirical Study." *Relations Industrielles* 52:4 (Fall 1997), 865–887.

MacLeod, Kathryn. *The Seniority Principle: Is it Discriminatory?* Kingston: Queen's University Industrial Relations Centre, 1987.

Mandel, Michael. *The Charter of Rights and the Legalization of Politics in Canada.* Toronto: Thompson Educational Publishing, 1994.

McDaniel, Susan. "Toward Disentangling Policy Implications of Economic and Demographic Changes in Canada's Aging Population." *Canadian Public Policy* 29:4 (2003), 491–509.

McDonald, Lynn P. and Richard A. Warner. *Retirement in Canada.* Toronto: Butterworths, 1990.

McDonald, Peter and Rebecca Kippen. "Labor Supply Prospects in 16 Developed Countries: 2000–2050." *Population and Development Review* 27 (2001), 1–32.

McIntosh, Tom. *The Post-Charter Labour Regime: The Trilogy and Beyond.* Kingston, Ontario: Industrial Relations Centre, Queen's University, 1989.

McLaren, Arlene T. and Margaret M. Manery. *Factors Affecting the Economic Status of Older Women in Canada: Implications for Mandatory Retirement.* Victoria: British Columbia Human Rights Commission, 2001.

McLaren, Arlene Tiger and Margaret Menton Manery. *Factors Affecting the Economic Status of Older Women in Canada: Implications for Mandatory Retirement.* Toronto: Canadian Centre for Policy Alternatives, 2001.

McMullin, Julie Ann. "Diversity and the State of Sociological Aging Theory." *The Geronologist* 40 (2000), 517–530.

McMullin, Julie Ann and Victor W. Marshall. "Ageism, Age Relations, and Garment Industry Work in Montreal." *The Gerontologist* 41:1 (2001), 111–22.

McPherson, Barry D. *Aging as a Social Process: An Introduction to Individual and Population Aging*, 2nd ed. Toronto: Butterworths, 1990.

McTavish, Donald G. "Perceptions of Old People." In David J. Mangen and Warren A. Peterson, eds., *Research Instruments in Social Gerontology: Clinical and Social Psychology*, Volume 1. Minneapolis: University of Minnesota Press, 1982, 533–621.

Michels, Robert. *Political Parties: A Sociological Study of the Oligarchical Tendencies of Modern Democracy*, trans. Eden and Cedar Paul. Glencoe: The Free Press, 1915.

Moen, Phyllis. "A Life-Course Approach to Retirement and Social Integration." In Karl Pillemer *et al.*, eds. *Social Integration in the Second Half of Life*. Baltimore: The Johns Hopkins University Press, 2000, 75–107.

Morton, Frederick L., ed. *Law, Politics, and the Judicial Process in Canada.* University of Calgary Press, 2002.

Myles, John. "Changing Public Policies, Changing Market Policies," in Peter Russell and Ken Rea, eds., *Redesigning Retirement: Proceedings of a Joint Forum Presented by the University of Toronto Faculty Association and the Retired Academics and Librarians at the University of Toronto: Innis Town Hall, Saturday 5 April 2003* (Toronto: RALUT, 2003), pp. 4–6.

Nascher, Ignatz L. *Geriatrics: The Diseases of Old Age and Their Treatment*, 2nd ed. Revised. London: Kegan Paul, French, Trubner, 1919 (orig. pub. 1914).

Necheles–Jansyn, Ruth F. "Retirement in Academe: Special Case or Social Model?" *Aging and Work* 6:3 (1983), 175–85.

Nelson, Todd D. ed. *Ageism: Stereotyping and Prejudice Against Older Persons.* Cambridge, MA: MIT Press, 2003.

Nelson, William H. *The Search for Faculty Power: the History of the University of Toronto Faculty Association, 1942–1992.* Toronto: UTFA, 1993.

Neumark, David. "Age Discrimination in the United States." In Zmira Hornstein, ed. *Outlawing Age Discrimination: Foreign Lessons, UK*

Choices. Bristol: The Policy Press, 2001.

Neumark, David and Wendy Stock. "Age Discrimination Laws and Labor Market Efficiency." *Journal of Political Economy* 107:5 (1999), 1081–125.

OECD. *Reforms for an Ageing Society.* Paris: Organisation for Economic Co-operation and Development, 2000.

Ontario Human Rights Commission. *Discrimination and Age: Human Rights Issues Facing Older Persons in Ontario.* Toronto, Ontario: Ontario Human Rights Commission, 2000.

Ontario Humans Rights Commission. *Policy on Discrimination Against Older Persons Because of Age.* Toronto, Ontario: Ontario Human Rights Commission, 2002.

Ontario Human Rights Commission. *Time for Action: Advancing Human Rights for Older Ontarians.* Toronto, Ontario: Ontario Human Rights Commission, 2001.

Ontario Task Force on Mandatory Retirement, R. Ianni, Chairman. *Fairness and Flexibility in Retiring from Work: Report of the Ontario Task Force on Mandatory Retirement.* Toronto: Ontario Ministry of Labour, 1987.

Osberg, Lars. "Is it Retirement or Unemployment? Induced Retirement and Constrained Labour Supply Among Older Workers." *Applied Economics* 25 (1993), 505–19.

Pesando, J.E. *The Elimination of Mandatory Retirement: An Economic Perspective.* Toronto: Ontario Economic Council, 1979.

Pesando, James and Morley Gunderson. "Retirement Incentives Contained in Occupational Pension Plans and their Implications for the Mandatory Retirement Debate." *Canadian Journal of Economics* 21:2 (1988), 244–64.

Pesando, James, Morley Gunderson and James McLaren. "Pension Benefits and Male-Female Wage Differentials." *Canadian Journal of Economics* 24:3 (1991), 536–50.

Phillipson, Chris. *Reconstructing Old Age: New Agendas in Social Theory and Practice.* London: Sage, 1998.

Policy of Research Initiative, Government of Canada. *Population Aging and Life-Course Flexibility: The Pivotal Role of Increased Choice in the Retirement Decision.* Discussion Paper. Ottawa: Government of Canada, 2004.

Prager, Joel. "Aging and Productivity: What Do We Know?" In David Cheal, ed. *Aging and Demographic Change in Canadian Context.* Toronto: University of Toronto Press, 2003, 133–89.

Preston, Samuel, Christine Himes, and Mitchell Eggers. "Demographic Conditions Responsible for Population Aging." *Demography* 26 (1989), 691–709.

Pupo, Norene and Ann Duffy. "Canadian Part-Time Work into the Millennium: On the Cusp of Change." *Community, Work and Family* 3:1 (2000), 49–69.

Pyper, Wendy and Philip Giles. "Approaching Retirement." *Perspectives on Labour and Income* 14 (2002), 9–16.

Quadragno, Jill, David MacPherson, Jennifer Reid Keene, and Lori Parham. "Downsizing and the Life-Course Consequences of Job Loss: The Effect of Age and Gender on Employment and Income Security." In Victor W. Marshall, Walter R. Heinz, Helga Kruger, and Anil Verma, eds. *Restructuring Work and the Life Course.* Toronto: University of Toronto Press, 2001, 303–18.

Reid, Frank. "Economic Aspects of Mandatory Retirement: The Canadian Experience." *Relations Industrielles/industrial relations* 43:1 (1988), 101–13.

Richter, Josef. "Economic Aspects of Ageing: A Review of the Literature." In George Stolnitz, ed. *Demographic Causes and Economic Consequences of Population Aging: Europe and North America.* New York: United Nations, 1992, 171–86.

Rifkin, Jeremy. *The End of Work*. New York: Tarcher/Putnam, 1995.

Rix, Sara E. "Restructuring Work in an Aging America: What Role for Public Policy?" In Marshall *et al.*, *Restructuring Work and the Life Course,* 375–96.

Robinson, Pauline K., Sally Coberly and Carolyn E. Paul. "Work and Retirement." In Robert H. Binstock and Ethel Shanas, eds. *Handbook of Aging and the Social Sciences*, 2nd ed. New York: Van Nostrand Reinold Company, 1985, 503–27.

Robertson, Ann. "'I Saw the Handwriting on the Wall': Shades of Meaning in Reasons for Early Retirement. *Journal of Aging Studies* 14:1 (2000), 63–79.

Rowe, Geoff and Huan Nguyen. "Older Workers and the Labour

Market." *Perspectives on Labour and Income* 3:2 (2003), 23–26.

Samorodov, Alexander. "Ageing and Labour Markets for Older Workers." *Employment and Training Papers Series, no. 33*. Geneva: International Labour Office, 1999.

Schellenberg, G. *Older Workers in the Canadian Labour Market.* Ottawa: The Canadian Seniors Network, 1994.

Schellenberg, Grant and Cynthia Silver. "You can't always get what you want: Retirement Preferences and Experiences." *Canadian Social Trends* 75 (2004), Statistics Canada, Cat. No. 11–008.

Schor, Juliet B. *The Overworked American: The Unexpected Decline of Leisure.* New York: Basic Books, 1991.

Scott, Tamra L. *Human Rights Issues and the Collective Agreement.* Kingston: Queen's University Industrial Relations Centre, 1996.

Seabrook, Jeremy. *A World Growing Old.* Toronto: Between the Lines, 2003.

Shannon, Michael and D. Grierson. "Mandatory Retirement and Older Worker Employment." *Canadian Journal of Economics/Révue canadienne d'Economique* 37:3 (2004), 528–51.

Singh, Gangaram and Anil Verma. "Is There Life after Career Employment? Labour Market Experience of Early Retirees." In Victor W. Marshall, Walter R. Heinz, Helga Krueger, and Anil Verma, eds. *Restructuring Work and the Life Course.* Toronto: University of Toronto Press, 2001.

Special Senate Committee on Retirement Age Policies. David A. Croll, Chairman. *Retirement Without Tears.* Hull, Quebec: Minister of Supply and Services Canada, 1979.

Spirduso, Waneen W. and Patricia Gilliam MacRae. "Motor Performance and Aging." In James E. Birren and K. Warner Schaie, eds. *Handbook of the Psychology of Aging*, 3rd ed. San Diego, CA: Academic Press, 1990, 184–200.

Statistics Canada. "Fact Sheet on Retirement." *Perspectives on Labour and Income* 4:9 (2003), 12–19.

Statistics Canada. "Fact Sheet on Retirement." *Perspectives on Labour and Income* 5:2 (February 2004), 1–22.

Statistics Canada. *General Social Survey: Social Support and Aging, 2002.* Cat. No. 89F0115XIE, Ottawa 2 September 2003.

Statistics Canada. *Labour Force Update: Older Workers.* Cat. No. 71–005–XPB. 1977.

Statistics Canada. *Permanent Layoffs, Quits and Hiring in the Canadian Economy: 1978–1995.* Cat. No. 71–539–XPB. Ottawa: Statistics Canada, 1998.

Statistics Canada. *A Portrait of Seniors in Canada.* Cat. No. 89–519–XPE. Ottawa: Statistics Canada, 1999.

Statistics Canada. "The Near–Retirement Rate." *Perspectives on Labour and Income,* Cat. No. 75–001–XPE (Spring 2004), 68–72.

Stebbins, Robert A. "The Extraprofessional Life: Leisure, Retirement and Unemployment." *Current Sociology* 48:1 (January 2000), 1–18.

Thomas, Marty and Rosemary A. Venne. "Work and Leisure: A Question of Balance." In David Cheal, ed. *Aging and Demographic Change in Canadian Context.* Toronto: University of Toronto Press, 2003.

Thorpe, Douglas. "Aging, Language and Culture" In Cheal, David. Aging and Demographic Change in Canadian Context. Toronto: University of Toronto Press, 2003.

U.S. Census Bureau. *Global Population Profile 2002.* U.S. Government Printing Office, Washington, D.C. 2004.

von Wachter, Till. *The End of Mandatory Retirement in the US: Effects on Retirement and Implicit Contracts.* Working Paper No. 49. University of California, Berkeley: Centre for Labor Economics, March 2002.

Waskett, Rosemary. "The Politics of Difference and Inclusiveness within the Canadian Labour Movement." *Economic and Industrial Democracy* 17 (1996), 587–625.

Young, Iris Marion. *Justice and the Politics of Difference.* Princeton: Princeton University Press, 1990.

Index